Science, Colonialism and Ireland

Science, Colonialism
and Ireland

Nicholas Whyte

CORK UNIVERSITY PRESS

Sci
Q
127
.I73
W49
1999

First published in 1999 by
Cork University Press
University College
Cork
Ireland

©Nicholas Whyte 1999

British Library Cataloguing in Publication Data
A CIP catalogue record for this book is available from the British Library
ISBN 1 85918 184 8 hardback
1 85918 185 6 paperback

Typeset by Tower Books, Ballincollig, Co. Cork
Printed by ColourBooks, Baldoyle, Co. Dublin

Contents

Preface

This book was initiated as a research project funded by the British Academy in the academic year 1991–2, to discover what archival materials might exist relating to the history of science in Ireland between 1890 and 1950. As such, it was intended to fill a number of gaps in the published history of science in Ireland. The first was a gap in time. Although Irish science and scientists of the nineteenth and earlier centuries had attracted a certain amount of academic attention from historians, and the more recent decades have come under the scrutiny of public policy studies, the period just before and during the first half of the twentieth century had been neglected. Part of my task was therefore to find out whether this neglect was due to a genuine decline of Irish science in the period or to some general scholarly bias. I have concluded that there are elements of truth to both explanations, though the actual decline of Irish science in the period is certainly the more influential of the two.

The second gap was that of a synoptic view of Irish science at any period. Published work on the whole had concentrated on particular scientific disciplines, institutions or individuals rather than trying to get a bigger picture. Although to get some idea of the bigger picture one inevitably must look at the smaller elements as well, I believe that I have developed a framework within which future research can be situated and which gives a convincing structure to the historical events that are recorded here.

A third gap was the general omission of Irish scientific writings from the canon of Irish culture. In the 1990s a number of voices have been pointing out this lacuna in the received version of Irishness. In particular the publishers of the *Irish Review* have lived up to their promise 'to publish articles on the arts, society, philosophy, history, politics, the environment and science', and both the Cultural Traditions Group of the Northern Ireland Community Relations Council and the Institute of Irish Studies at the Queen's University of Belfast have supported the exploration of the history of science. My own work was supported in the academic year 1995–6 by the Cultural Traditions Group, who awarded me the Mary Ward Junior Fellowship at the Institute of Irish Studies.

Professor Peter J. Bowler, the originator of the British Academy project and my supervisor for the doctoral thesis on which this book is based, was always ready to discuss the work and the wider issues within the history of science that it impacted. I have heard of no other supervisor who has organised an entire conference (in Armagh, in 1994) devoted to a graduate student's subject when that subject is quite far removed from his or her own field of interest. He and Sheila Bowler, Robert E. Hall, and Valeria Lima Passos formed the social and collegiate background which is essential to support any research in any discipline, backed by the other staff and students of the School of Philosophical and Anthropological Studies at Queen's University. The British Academy's involvement in the original research project has already been acknowledged. I also received a travel grant from the Royal Society at that time. For my three years as a full-time research student I was supported by a state studentship from the Department of Education for Northern Ireland. The directors of the Cultural Traditions Group and of the Institute of Irish Studies, Dr Maurna Crozier, and Dr Brian Walker, provided much-needed advice and support. I am only sorry that the project envisaged for the Mary Ward Fellowship did not emerge in the form originally intended.

The librarians and archivists of the following institutions made research an enjoyable experience: Cambridge University Library; the National Archives, Dublin; National Library of Ireland; National Museum, Dublin; Edinburgh University Library; the Bodleian

Library, Oxford University; the Linenhall Library, Belfast; the Plunkett Foundation for Co-Operative Studies, Oxford; the Public Records Office, Kew; the Royal Botanic Gardens, Kew; the Royal Dublin Society; the Royal Irish Academy; Trinity College, Dublin; University College, Cork; University College, Dublin; University College, Galway; and University College, London. Particular thanks are due to Graham McKenna of the British Geological Survey, Keyworth, Nottinghamshire; the late John Thackray of the Natural History Museum, South Kensington; Jean Archer of the Geological Survey of Ireland; and Jim O'Connell of the Natural History Museum, Dublin. The staff of Queen's University library, particularly those in the special collections department, have been unfailingly helpful. I am grateful to the relevant bodies including the Board of Trinity College Dublin and the Council of trustees of the National Library of Ireland, for access to and permission to publish from their manuscript collections.

The following individuals helped me via conversation, correspondence, insights or anecdotes: Hermann and Mary Brück; Barbara, Lady Dainton; Gordon Herries Davies; Sean Faughnan; Tom Garvin; Richard Jarrell; Sean Lysaght and an anonymous publisher's reader; John McConnell; Roy McLeod; Louis McRedmond; Maurice Manning; Patrick Maume; Christopher Moriarty; the late Ernest Walton; Trevor West; and Jim White. Conor Kostick, Emm Barnes and Sam Inglis helped obtain some crucial snippets of information, and Dennis Burton reviewed the whole text at a late stage. Needless to say, any remaining errors are entirely my responsibility.

The electorate of North Belfast did not support me in sufficient numbers in May 1996 to elect me as a representative to the 1996–8 Northern Ireland Forum and multi-party talks, and so I was able to complete my thesis. For some reason I feel more grateful towards the 1,670 who did vote for me. More seriously, so many members of the Alliance Party of Northern Ireland have provided moral support for my academic as well as my political work over the last few years that it would be invidious to single out any individual, apart from Garth Gilmour who provided valuable feedback and proof-reading on several chapters.

Since I completed my thesis, my career has shifted to the field of democratisation in the former Yugoslavia. This may seem a remote departure from the history of science in Ireland, but in fact I find that there are some relevant universals: the nature of divided politics, the ability of any society to recover from devastating conflict, and the importance of historical truth. Despite my distant and migratory lifestyle, Sara Wilbourne and colleagues at Cork University Press have kept in touch and ensured that we remain on track.

My mother, Dr Jean Whyte, and my brother and sister, William and Caroline, provided much-needed support at numerous stages of the work; my daughter Bridget arrived to delight us all only during its final stages. But my chief debt of gratitude for providing a background in which the work was possible is owed to my wife Anne, who also provided the necessary voice of common sense on many occasions. It is to her that this book is dedicated.

Nicholas Whyte
Zagreb, November 1998

1. Introduction

Irish science, 1890–1930, and Irish cultural studies

This book is a study of the social context of science in Ireland during one of the more interesting periods of recent Irish political history: the decades leading up to and immediately after Partition and the independence of the Irish Free State. Science in Ireland, for present purposes, is largely restricted to the academic disciplines pursued in institutions of higher education, government research institutes, and industrial research, and co-ordinated through such bodies as the Royal Dublin Society and the Royal Irish Academy. Medicine and engineering have had to be excluded here. Both are potentially substantial areas of research in themselves, and restricting the field of immediate interest to astronomy, chemistry, physics, geology, zoology and botany has produced material enough.

The subject of this book is the intersection over one period of time of the history of science and Irish studies, two disciplines which have developed in almost total independence of each other. Although much Irish history is still the history of Irish politics – and sadly it is also true that most Irish politics has been the politics of Irish history – the history of science has begun to find a niche within Irish cultural studies. A number of recent articles in the *Irish Review* and *Éire-Ireland* by Roy Johnston, Dorinda Outram, Sean Lysaght, John Wilson Foster, Greta Jones and the present writer have addressed historical and cultural aspects of science in

1

Ireland.[1] It is interesting that Lysaght and Foster are both lecturers in English literature rather than in history. Few recent contributions on the history of science in Ireland have appeared in any Irish history journal.[2]

With this trend has come a recognition that the interaction between Irish culture and the natural world has had a profound influence on many aspects of Irish life, including economic development, environmental politics, health and medical issues and the feminist agenda. Lysaght rightly complains that in academic Irish history, as it has been constructed over the last fifty years, 'science is conventionally ignored, or assigned marginal status either as a component of Victorian culture, or as a very recent arrival into debates about educational policy'.[3] But while this may be true of the canon of Irish history, it is no longer so for Irish cultural studies. An international conference was held in Armagh in 1994 to address the social context of science, technology and medicine in Ireland since 1800. Significantly, it received funding from the Cultural Traditions Group of the Northern Ireland Community Relations Council.

It cannot be denied that Irish history as a discipline has ignored science. John Wilson Foster has examined a number of standard histories of Ireland and Irish studies, and has drawn attention to the

---

1   Dorinda Outram, 'Negating the Natural: Or Why Historians Deny Irish Science', *Irish Review* 1 (1986), 45–9; John Wilson Foster, 'Natural History, Science and Irish Culture', *Irish Review* 9 (1990), 61–9, and 'Natural Science and Irish Culture', *Éire-Ireland* 26, no. 2 (1991), 92–103; Nicholas Whyte, '"Lords of Ether and of Light": The Irish Astronomical Tradition of the Nineteenth Century', *Irish Review* 17/18 (1995), 127–41; Roy Johnston, 'Science and Technology in Irish National Culture', *Crane Bag* 7 (1983), 58–63, 'Science in a Post-Colonial Culture', *Irish Review* 8, 70–6, and 'Godless Colleges and Non-Persons', *Causeway* 1, no. 1 (Autumn 1993), 36–8; Sean Lysaght, 'Heaney vs. Praeger: Contrasting Natures', *Irish Review* 7 (1989), 68–74, and 'Themes in the Irish History of Science', *Irish Review* 19 (1996), 87–97; Michael Viney, 'Woodcock for a Farthing: The Irish Experience of Nature', *Irish Review* 1 (1986), 58–64.

2   Richard Jarrell, 'The Department of Science and Art and Control of Irish Science, 1853–1905', *Irish Historical Studies* 23 (1983), 330–47, and Greta Jones, 'Eugenics in Ireland – The Belfast Eugenics Society, 1911–1915', *Irish Historical Studies* 28 (1992), 81–95.

3   In his Ph.D. thesis, 'Robert Lloyd Praeger and the Culture of Science in Ireland: 1865–1953' (NUI 1994).

lack of any discussion of science in any of them.[4] Gordon Herries Davies quips that 'our historians have felt more comfortable in discussions of banking, battles and bishops than in dealing with problems concerning basalt, binomials and brachiopods'.[5] But the history of science is not unique in this respect. Irish history as a discipline has also traditionally ignored the history of women and the history of the labour movement, omissions which have begun to be rectified in recent years. The present work is a contribution to filling that gap in information.

Historians of science have not been as neglectful of the Irish dimension to their subject. Most history of science naturally enough concerns science in the world's dominant economies (the USA, France, Britain, Germany, and more recently Russia and Japan). Smaller countries present different problems. The recent run of case-by-case studies of particular Irish personalities, Irish institutions, or scientific disciplines in Ireland, usually written by scientists (active or retired) rather than historians, could be matched by a number of other European nations.[6]

Most small European nations or regions have been under the domination of at least two and often more of the great powers in the last 200 years. Ireland is unusual in that its relationship to external countries was until very recently affected by its relationship with just one of its neighbours, Britain. There has been considerable debate about the extent to which this relationship can be described as a colonial one. At the time of independence Irish Nationalists did

---

4   Foster, 'Natural Science and Irish Culture', p. 92. The works in question are
    Terence Brown's *Ireland: A Social and Cultural History*, Roy Foster's *Modern Ireland 1600–1972*, F.S.L. Lyons' *Ireland since the Famine*, and Joe Lee's *Ireland 1912–1985*.
5   Herries Davies, 'Irish Thought in Science', in Kearney, *The Irish Mind* (1985), p. 310.
6   One of the most notable is Finland, where the Finnish Academy of Sciences has produced a number of monographs on the history of various branches of learning in Finland in the 1828–1918 period, including G. Elfving, *The History of Mathematics in Finland, 1828–1918* (1981), R. Collander (tr. by David Barrett), *The History of Botany in Finland, 1828–1918* (1965), p. 107–11, and T. Enkvist, *The History of Chemistry in Finland, 1828–1918* (1972). I also found useful a brief history of science in the Netherlands: Klaas van Berkel, *In het voetspoor van Stevin: geschiedenis van de natuurwetenschap in Nederland 1580–1940* (1985).

not relish comparison with British colonies overseas. The suggestion that the Northern Ireland crisis is a colonial situation dates only
from 1970.[7] More recently, Liam Kennedy has pointed out the lack
of congruence between the Irish and colonial conditions.[8] The term
'colonial' has acquired a sharply political resonance in Irish studies,
implying as it does both the illegitimacy and the impermanence of
the colonial regime. In particular, the idea of applying a colonial
analysis to Irish history has become discredited by association
among historians in recent years because of its adoption by the
republican movement. Michael Hechter's exploration of Ireland and
the rest of Britain's Celtic periphery as an 'internal colony', rather
than a classic colonial situation, has not yet proved to be a successful *via media* for Irish historians.[9]

In the field of Irish literary studies, on the other hand, the idea that
Ireland is a 'post-colonial' nation, and that this is necessarily reflected
in its culture, has become increasingly popular. In particular, Declan
Kiberd has argued that Irish history must be seen in terms of other
decolonising nation-states around the world, rather than in comparison with the major European or Anglophone states. As science is in
many ways a cultural practice, the history of Irish science is an
important test case for Kiberd's thesis.[10]

History of science too has seen a recent increase of interest in the
development of science in societies which were, and arguably still
are, under the cultural and economic hegemony of the imperialist
powers. It is convenient to call this 'colonial science' or 'imperial
science'. The former seems to have become the more popular
description in the discipline, although the term 'colonial' is not
always literally accurate. Ottoman Turkey, for instance, was never
formally colonised by anyone except the Turks themselves, and the
relationship between France and Brazil in the mid-nineteenth
century was formally between two empires.

7   J.H. Whyte, *Interpreting Northern Ireland* (1990), pp. 177–8.
8   Kennedy, *Colonialism, Religion and Nationalism in Ireland* (1996); also 'Modern
    Ireland: Post-Colonial Society or Post-Colonial Pretensions?', *Irish Review* 13
    (Winter 1992/93), 107–21.
9   Hechter, *Internal Colonialism: The Celtic Fringe in British National Development,
    1536–1966* (1975).
10 I am summarising Kiberd's views from Joe Cleary's review article, 'Irish
    Culture as Decolonized Nation-State', *Irish Literary Supplement* (Fall 1996), 19–20.

It is my intention therefore to assess the social context of science in Ireland in the closing years of British rule there in the light of what has been learnt from situations of colonial science, to offer a framework to better understand the actions of scientists in Ireland during this period, and to assess the extent to which the Irish context can really be referred to as a 'colonial' one.

Irish science and colonial science

There has not been an overwhelming amount of historical research on Irish science in this period. Gordon Herries Davies' 1985 bibliographical handbook for the history of Irish science remains an essential starting point. Other important resources in the field published since then which include our period include institutional histories of the Royal Irish Academy, Dunsink Observatory, the Botanic Gardens at Glasnevin, the Armagh Observatory, the Geological Survey of Ireland and University College Cork; a survey of the Irish scientific instrument trade up to 1921; biographical studies of William Sealy Gosset and Robert Lloyd Praeger; and several collections of biographical sketches of Irish scientists for the educational market. There have been at least three recent conferences on the history of science in Ireland. The papers given at the first (at the Royal Irish Academy in 1985) have not been published as a unit, but those from the second and third (respectively at Trinity College Dublin in 1988 and the Royal School, Armagh, in 1994) have been, and another recent volume brings together a number of writers on natural history and Irish culture.[11]

A seminal overview of the subject is Gordon Herries Davies' 1985 essay on 'Irish Thought in Science'. This pointed out the comparative success of science in Ireland throughout the nineteenth century,

---

11 Herries Davies, 'The History of Irish Science: A Select Bibliography', second edition (duplicated typescript, 1985); T. Ó Raifeartaigh (ed.), *The Royal Irish Academy: A Bicentennial History, 1785–1985* (1985); P.A. Wayman, *Dunsink Observatory, 1785–1985: A Bicentennial History* (1987); E. Charles Nelson, *The Brightest Jewel: A History of the National Botanic Gardens, Glasnevin, Dublin* (1987); J.A. Bennett, *Church, State and Astronomy in Ireland: 200 Years of Armagh Observatory* (1990); Gordon L. Herries Davies, *Sheets of Many Colours: The Mapping of Ireland's Rocks, 1750–1890* (1983), and *North from the Hook: 150 Years of the Geological Survey of Ireland* (1995); J.A. Murphy, *The College: A History of Queen's University College*

and its apparent decline after about 1890, and also raised the issues of the comparative dearth of Catholics among Irish scientists of that period and the absence of the history of science from Irish history in general.[12] The British Academy-funded project which developed into the writing of this book was specifically intended to investigate Irish science at the time of its apparent decline, from 1890 to the granting of independence to the Irish Free State in 1922.

The apparent lack of connection between, on the one hand, the scientific tradition in Ireland, and, on the other, the Catholic church and the politics of Irish nationalism, is indeed striking. Roy Johnston and John Wilson Foster, among others, conclude that the Catholic church itself discouraged science among its followers, and that the culture of Irish nationalism as it developed, being anti-modern and anti-Protestant as well as nationalist, must have been responsible for the failure of the Irish state to utilise its (mainly Protestant) scientific potential after independence. As Foster puts it, 'Science itself was in danger of being ethnicised as foreign (not truly Irish), categorised as anti-religious (not acceptably Catholic) and classified as socially distant.'

This book is aligned with the work of Steven Yearley, Richard Jarrell, Jim Bennett and others, to reveal a more complex story. There is less to the case against either the church or nationalism than

---

11 (*cont.*)   *Cork, 1845–1995* (1995); J.E. Burnett and A.D. Morrison-Low, *Vulgar and Mechanick: The Scientific Instrument Trade in Ireland, 1650–1921* (1989); 'Student', a *Statistical Biography of William Sealy Gosset* (1990); Sean Lysaght, 'Robert Lloyd Praeger and the Culture of Science in Ireland: 1865–1953' (Ph.D. thesis, NUI 1994); Charles Mollan, William Davis and Brendan Finucane, eds, *Some People and Places in Irish Science and Technology* (1985); Charles Mollan, William Davis and Brendan Finucane, eds, *More People and Places in Irish Science and Technology* (1990); Wilbert Garvin and Des O'Rawe, *Northern Ireland Scientists and Inventors* (1993); the Blackstaff Primary Science Key Stage 2 textbooks (1995); Susan McKenna-Lawlor, *Whatever Shines Should be Observed [quicquid nited notandum]* (1998); J.R. Nudds, N.D. McMillan, D.L. Weaire and S.M.P. McKenna-Lawlor, eds, *Science in Ireland, 1800–1930, Tradition and Reform, proceedings of an international symposium held at Trinity College, Dublin, March, 1988* (1988); Peter J. Bowler and Nicholas Whyte, eds, *Science and Society in Ireland* (1997); Greta Jones and Elizabeth Malcolm, eds, *Medicine, Disease and the State in Ireland, 1650–1940* (1998); John Wilson Foster, ed., *Nature in Ireland: A Scientific and Cultural History* (1998).

12 Herries Davies, 'Irish Thought in Science', in Kearney, *The Irish Mind* (1985), pp. 294–310.

might have been expected. There is also a strong case to be made that the mainly Protestant scientific elite at least colluded in the exclusion of Catholics from the scientific mainstream. In the divided society that was Ireland of this period, both sides were to blame for the perpetuation of the division.

Three strands in Irish science

A new interpretative framework for the social context of Irish science between 1890 and 1930 is proposed here. This framework has its roots in the models of the development of colonial science proposed for the cases of America by George Basalla, and for India by V.V. Krishna, but amends them and introduces a number of features unique to Ireland. It is not a prescriptive model for the growth of science in other societies, but it does point to possible future areas of cross-cultural comparison and of the further development of the history of Irish science.

The first model for the spread of Western science was that put forward by George Basalla in 1967.[13] Basalla proposed a three-phase process for the diffusion of scientific practice into non-Western nations. The first phase is that of the scientific exploration of a new territory as a resource for European science; in the second phase, 'colonial science', scientific work is carried out by residents of the territory in question who see themselves as dependent on the scientific tradition of the metropolis; and in the third phase, the colony develops its own national independent scientific tradition. Each of the first two phases peaks and declines in activity in turn, while the third will continue to expand. All three may in fact coexist at any one point in time.

Basalla's model has not been widely applied to countries other than the US, but it has prompted a number of other writers to put forward their own alternatives. A majority of the essays in two recent volumes on colonial/imperial science begin by defining their position with respect to Basalla.[14] To take one example, Roy MacLeod points

---

13 G. Basalla, 'The Spread of Western Science', *Science* 156 (1967), 611–22.
14 *Science and Empires: Historical Studies about Scientific Development and European Expansion*, ed. Patrick Petitjean, Catherine Jami and Anne-Marie Moulin (1992); and *Scientific Colonialism: A Cross-Cultural Comparison (Papers from a Conference at Melbourne, Australia, 25–30 May 1981)*, ed. Nathan Reingold and Marc Rothenberg (1987).

out that the Basalla model is too linear and homogeneous with respect to both the process of transmission and the nature of the scientific enterprise itself. He suggests that instead of regarding the relationship between the colony and the original metropolis, we should look at the extent to which the metropolis was mobile. Different scientists at different times might validate their work with reference to the imperial capital, to some more local venue, or to some other location entirely. MacLeod goes on to put forward his own framework, a five-stage taxonomy of British imperial science between 1780 and 1939, and makes only modest claims for its more universal applicability.[15]

MacLeod largely addresses the 'settler' colonies of Australia, New Zealand and Canada, where the national science tradition that developed was that of the descendants of Europeans and not of the aboriginal inhabitants. In both India and Ireland, by contrast, there had been a developed and industrialising economy before 1800, and British rulers always had to negotiate carefully with 'native' elites. V.V. Krishna, in an essay on the emergence of 'national science' in India between 1876 and 1920,[16] classifies the scientists who were then active in India into three broad categories. Although Krishna explicitly rejects Basalla's three-phase model, the parallels between his categories of 'colonial scientist' and the earlier writer's 'phases' are clear.

Krishna's model categorises scientists by their role within the larger story of imperialism and nationalism. His first category is the 'gate-keepers', those whose sympathies were with the British imperial ideal, actively trying to exclude the 'natives' from scientific careers; he singles out the Indian Geological Survey as a gatekeeping institution. The second group are the 'scientific soldiers', Indian or European by race, who were trained by Britain and sent out to do lower-status but professional scientific jobs in the Empire; he names two railway engineers by way of example. In the third category are

---

15   MacLeod, 'Reflections on the Architecture of Imperial Science', in Reingold and Rothenberg, *Scientific Colonialism*, pp. 217–49. MacLeod's essay 'On Science and Colonialism', in Bowler and Whyte, *Science and Society in Ireland,* helpfully puts Ireland in an international context.

16   V.V. Krishna, 'The Colonial "Model" and the Emergence of National Science in India: 1876–1920', in P. Petitjean et al., *Science and Empires* (1992), pp. 57–72.

those scientists such as J.C. Bose and C.V. Raman responsible for the growth of a local, independent scientific culture, whom Krishna hesitantly identifies as nationalists. Krishna points out that:

> explicit to [this] stratification is also the recognition that scientists in each stratum had different constituencies of operation with different goals and purposes. It is held that the third category was responsible for the emergence of national science or an 'independent' scientific phase and that the other two categories were not.

Ireland of course is not India, but parallels can be drawn between Krishna's categories and three strands or traditions within Irish science of the same period. The 'gatekeepers' in the Irish context are clearly the scientists of the Ascendancy, who had a history of excluding those from outside their own tradition during the nineteenth century. In the early twentieth century these institutions had to make accommodations with the new Ireland, and as we shall see some were more successful than others. A second strand within Irish science comprises the scientific civil servants, whose numbers grew in Ireland up to 1900 and for a short time thereafter. Although science has no nation, nations do have science, and the scientists of the administration had to negotiate the claims of conflicting nationalities. Finally, the scientists of the nationalist tradition in India have a clear parallel in those Irish scientists who were supporters, in some cases very prominent supporters, of the various nationalist and separatist movements in Ireland.

Ireland's Gatekeepers: the Ascendancy

A substantial proportion of Ireland's population not only approved of British rule, but were considered by the British as an integral part of the British state. Even though direct political control of Ireland had been taken out of the hands of the Protestant elite in 1801, they remained economically and educationally dominant in the island as a whole throughout the nineteenth century. It is tempting to equate the British administrators of India with the landlords of the Irish Ascendancy, particularly given the resistance displayed by both groups to any increase in power for the Indians in the first case or for Irish Catholics in the second. But the analogy is incomplete, because the Irish elite were in fact Irish themselves. A considerable majority of the Irish scientists of the nineteenth century came from the Ascendancy tradition; various surveys have found only between 8 per cent and 14 per cent who were Catholics.

Part I examines several aspects of Ascendancy science, with particular reference to Steve Yearley's suggestion that science in Ireland was primarily 'practised and valued as a cultural activity. Scientific attainment might be esteemed but it would be evaluated alongside other cultural accomplishments, not in terms of its utility.'[17] One example of a scientific pursuit with little obvious practical utility but which was brought to unprecedented levels in nineteenth-century Ireland was astronomy. By the end of the century the tradition was dying out, and this decline has recently been attributed to political divisions among Irish astronomers along the Home Rule/unionism cleavage. It seems however more likely that changes in land ownership and the increasing cost of such projects had weakened the Ascendancy's ability to carry them through, and that the professionalisation of astronomy had made it a much more difficult discipline for the amateur to break into, no matter how rich.

The Royal Dublin Society and Trinity College Dublin are obvious examples of institutions whose interest was in maintaining the dominance of a scientific elite. Elizabeth I had founded Trinity College, in the centre of Dublin, in 1592 in order to spread Protestantism among the Irish. Although it had abolished formal religious discrimination in 1869, it maintained a firm commitment to the Union and an equally firm Ascendancy ethos. This was reinforced by the Catholic church strongly discouraging its members from sending their sons (and from 1903 their daughters) to the college, lest their faith and morals be endangered. The college succeeded in avoiding institutional change during the resolution of the vexed issue of providing a higher-education system acceptable to Catholics. However, its financial problems came to a head after the founding of the Irish Free State in 1922, and some previously unknown details of the negotiations which temporarily resolved that problem will be described.

---

17   Steve Yearley, 'Colonial Science and Dependent Development: The Case of the Irish Experience', *Sociological Review* 37 (1989), p. 319

The Royal Dublin Society had been involved with the promotion of scientific research since its foundation in 1731, and many of its scientific initiatives had been taken over by the British government in the nineteenth century. Even until recently it published several scientific journals and held regular scientific lectures at its headquarters, Leinster House, in the centre of Dublin. As its main purpose was agricultural, and many members were from the landowning classes, it is not surprising that the RDS was very much identified with the Ascendancy. As late as 1925, it is clear that Catholic scientists seeking election to the RDS's Committee for Science and its Industrial Applications were much less likely to be elected than were Irish Protestants.

Although neither the RDS nor Trinity College (after 1869) practised formal discrimination against Irish Catholics, it is difficult to see how Catholic or nationalist scientists could have felt comfortable working within either institution. Some scientists associated with both had extreme views. George Francis Fitzgerald, fellow of TCD and secretary of the RDS, writes of his disgust with Papists to his friend Oliver Lodge in 1891, and even a sympathetic writer describes his colleague John Joly, professor of geology in Trinity and president of the RDS in its bicentennial year, as 'blind and deaf to nationalist sentiment which he called lawlessness and sedition'.[18]

The Royal Irish Academy, then as now the internationally recognised institution for Irish science, might also have been expected to play a gatekeeping role, with its strong historical links with Trinity College and its royal patronage. In fact it seems to have become politically balanced between the three groups of Irish scientists early enough to ensure its own survival. This may have come about because of the academy's dual function of promoting both arts and sciences. Its catalytic role in the genesis of Celtic studies in the mid-nineteenth century had attracted a substantial number of nationalists into its previously Ascendancy-dominated membership. The academy's most substantial scientific project in the years before independence, the Clare Island Survey, includes a chapter by the nationalist academic Eoin MacNeill on Gaelic terms used on the

18 Terence de Vere White, *The Story of the Royal Dublin Society* (1955).

island. The scientific members of its committee, elected by the members at large, were drawn from all three groups, with perhaps a bias towards the Ascendancy before independence, which did not survive the traumas of 1920–22. This history of pluralism equipped the RIA to survive a challenge to its status as the national institution of Irish learning by a group of nationalist scholars.

The RIA's adaptation to the new order after independence is the success story of the gatekeepers. Trinity College had been guaranteed a substantial income by the British government as long as it remained in Ireland, but this agreement lapsed when the British withdrew, and Trinity became a largely self-funded institution, drawing on its reserves of past glory. Its lack of capital resources meant that new spending on scientific activities was very difficult. It was not ignored by the new state – its graduates continued to elect members to the Dáil, and still retain parliamentary representation – but it was not until the Catholic church's ban was withdrawn in the 1960s that it could be said to have fully taken its place in modern Ireland. The Royal Dublin Society, led by one innovative administrator, adapted much more quickly by getting a large cash settlement from the new government, which had taken over the RDS headquarters. It then widened its membership, drastically cut back its scientific activities and concentrated on the organisation of popular agricultural events at which it has remained so successful.

Administrators

The British government in Ireland[19] employed many scientists. Some were English in origin, some Irish Protestants, a very few were Catholic, and most could have been called middle-class. As employees of the state, their jobs were fairly secure even under the threat of a Home Rule government, which few of them supported. As with Krishna's 'scientific soldiers' in India, they were not as socially elevated as the Ascendancy scientists but were still associated with the British state which paid their salaries. Of the three groups considered here, they were perhaps the most 'colonial' in that they were least likely to be firmly committed to any political view.

---

19 Referred to (perfectly accurately) by its contemporaries as the Irish government.

Most state scientists were employed by one key body: the Department of Agriculture and Technical Instruction for Ireland. DATI, as it was generally known, was effectively a ministry for agricultural development, set up as a result of successful lobbying by Sir Horace Plunkett, a Unionist MP who became its first vice-president (the department's nominal president was the chief secretary for Ireland). Plunkett's lobbying in favour of the creation of DATI succeeded partly because he was able to unite an impressive coalition of businessmen and Unionist and Nationalist politicians behind his campaign, but far more so because it fitted in extremely well both with the Irish policy of the British government and with its attitude to the state's role in science. Plunkett's tenure of the vice-presidency survived the change of government in 1905 by two years and the loss of his own parliamentary seat in 1900 by seven because he had succeeded in making the work of the department 'non-political' and in making himself appear indispensable to its continued operation.

On its foundation in 1900, the Department of Agriculture and Technical Instruction inherited a number of Irish institutions which had been brought under the umbrella of the Department of Science and Art over the previous century. These included the National Museum, the Royal College of Science, the department's Fisheries Branch, and from 1905 the Geological Survey of Ireland. These institutions had originated as separate bodies, in most cases under the sponsorship of the RDS, but were now being administered as a group by the new department as instruments for the general improvement of Irish science.

The 'administration scientists' employed in these institutions had all joined while they were part of a centralised imperial administration run from South Kensington. Most of them were English and were committed to the British Empire as a natural framework for the co-ordination of scientific effort. Their reactions to the changing nature of the relationship between Dublin and London modified the strategies they adopted to survive within their environment and to increase the perceived importance of their own areas of expertise. Three case studies are examined in which English-born scientists in the Irish administration were prepared to use nationalist political considerations as a weapon in their professional disputes. These cases support Roy MacLeod's suggestion that a colonial scientist had to work against the background of a 'moving metropolis'.

The administration scientists had mixed fortunes after independence. Their ranks had been somewhat depleted by the First World War: some had been killed, others had reaped the benefits of having a scientific training in wartime and moved to jobs in England. The 1921 peace settlement had included a clause allowing those who disagreed with the new government to leave their jobs but retain their pension rights, and some used this opportunity to leave Ireland (others treated it as a form of early retirement and stayed). The unwillingness of the new government to spend any money at all – a well-known characteristic of the Irish Free State's Department of Finance in the decades immediately following independence – certainly extended to science, and the depletion of the Geological Survey is unusual only in that it is comparatively well chronicled. The staff of the Royal College of Science discovered to their indignation that they were to be transferred to University College Dublin, and the story of its closure in the midst of the 1922–3 civil war is told here in full for the first time.

Nationalists

In the history of Irish science, the contributions of the nationalist and Catholic tradition have been least researched, although there had always been a number of Catholic scientists, even in the nineteenth century. The provision of higher education for Catholics on terms acceptable to their church was one of the few Irish issues which was capable of raising political problems in England. Although this state of affairs was gradually easing at the start of the twentieth century, it was not resolved until the foundation of the National University of Ireland in 1908. The National University was non-denominational on paper, but its governing bodies were appointed to ensure that it would be Catholic and nationalist in ethos, and that new appointments would on the whole follow this trend. It was in the National University that the scientists who would serve the new nation were trained, and so it was the National University rather than Trinity College or the state-employed scientists which was enabled to define what science would become in the new Ireland.

There is very little evidence to suggest that the Catholic church was discouraging of the practice of science in nineteenth-century Ireland. The only known case of a Catholic scholar suffering ecclesiastical censorship was theological rather than scientific, and Catholics wrote on all aspects of science, including evolution and geology, during the period. It is much more convincing to attribute the lack of Catholic representation among nineteenth-century scientists to the failure of the state to provide a system of higher education acceptable to Catholics and the inability of the church to raise enough money to adequately fund its own university in Dublin.

Similarly, the case against Irish nationalism as an impediment to science is far from proven. While it is true that much nationalist ideology was romantic and looked to a rural idyllic and Gaelicised past, prominent figures in nationalist politics such as Arthur Griffith, D.P. Moran, Arthur Lynch, Darrell Figgis, Timothy Corcoran, Alfred O'Rahilly and Eamon de Valera all regarded science and/or industry to a certain extent as important parts of their visions of state-building.

Other areas of interest

The three-stranded framework outlined above is one in which most Irish scientific activity of this period can be placed. There are other aspects of the history of Irish science which could have been explored, time and space permitting, and three of those which might have been included are outlined below as signposts for future research.

Women in Irish Science

In the context of the current debates about women in the history of science and women in Irish history, it will be very important at some future stage to investigate how women scientists were treated in Ireland. The initial impression is that they were treated differently by each of the three strands. Women could not be part of a male scientific elite, and did not gain any recognition from the gatekeepers. No women studied at Trinity College at all until 1903, and none were scientifically active within the RDS or RIA until the middle of this century. Administration science did not exclude women, but scientists such as Mabel Wright, Anne Massie, Matilda Knowles and Jane Stephens (later Scharff) might survive for decades on temporary contracts without gaining recognition from their scientific peers, and would be dismissed from government jobs on marriage. They were allowed some participation in the natural history movement, but the (male) professional scientists set the agenda. While the National University seems to have offered most hope of advancement for women scientists – Hugh Ryan, the professor of chemistry at UCD, published papers jointly with a number of women between 1915 and 1930 – even this was not very great. Married women in particular were discouraged from working until quite recently.

Belfast: A Place Apart

I have not yet mentioned Northern Ireland or the Queen's University of Belfast, which was elevated to separate university status when the National University was founded in 1908. Its exclusion from the NUI demonstrated that it was not expected to form part of the Irish nation on the same terms as Dublin, Cork and Galway, and its omission from this study could be excused on the grounds that it did not end up as part of the Irish Free State after 1922. A full analysis of QUB's location in cultural space would require a study in itself: its special position in Ireland is influenced by the greater industrialisation of Ulster and also by its geographical and cultural links with lowland Scotland. I stated earlier that Ireland could not be treated simply as a regional case of British science. Belfast however is a regional case rather than a national one, needing to be located at the intersection of Ireland and Scotland, within the general context of metropolitan British science, in a way that the rest of Ireland is not.

Natural History and Quaternary Studies

The natural history movement was one of the most successful branches of science in Ireland at the start of the twentieth century. Like astronomy, it had always been a discipline of amateurs, but it was a pursuit of the middle classes, not the rich. Co-ordinated by people like Robert Lloyd Praeger and supported by his National Museum colleagues, the natural history field clubs survived the change of government battered but intact. Although the *Irish Naturalist*, published in Dublin, folded in 1924, it was immediately replaced by the *Irish Naturalists' Journal*, published from Belfast but with very similar content (and contributors). Natural history is clearly the amateur wing of the administration tradition of science in Ireland, and the work of natural historians was usually validated through the National Museum and the other government scientific institutions. Its very importance has effected some excellent historical research on the subject in the last few years.

Although natural history was not completely professionalised, it had a professional backbone and so was able not only to endure the independence struggle but also to beget one of independent Ireland's scientific innovations, quaternary studies. In 1933 the Committee for Quaternary Research in Ireland was set up by Robert Lloyd Praeger, J.K. Charlesworth of Queen's University of Belfast, Anthony Farrington of the Royal Irish Academy and Henry Seymour of UCD (both these last two were ex-Geological Survey). The committee invited a Danish geologist, Knud Jessen, to spend 1934 in Ireland, and a whole new field of Irish science opened up, based not on British models but on the common experience of another small (and mostly flat) northern European country. Perhaps this was the beginning of a new synthesis of the three strands of Irish science.

Part I

2

The Ascendancy scientists

The intellectual and political tradition which draws its roots from the eighteenth-century Ascendancy in southern Ireland is often described as 'Anglo-Irish'. That term will not be used here. Apart from anything else, it is ambiguous. A search of the 200 or so works in the Queen's University of Belfast's library catalogue which contain the word 'Anglo-Irish' in their titles reveals that only about half of them use the term in that sense. The most frequently used – and reasonable – alternative meaning of 'Anglo-Irish' is 'involving the interests of both England/Britain and Ireland' as in the 1985 'Anglo-Irish Agreement' and the 1921 'Anglo-Irish Treaty'. Other meanings include 'the dialect of the English language spoken in Ireland', and a classification of archaeological artefact usually used in opposition to 'Hiberno-Norse'.

The use of the phrase 'Anglo-Irish' incorrectly implies an 'English' component to the identity of the Ascendancy tradition. In fact the Ascendancy were just as distrustful of their English rulers as were Irish Catholics, and with better reason, for the history of Ireland under the Union is the story of one democratic or social reform after another, each systematically chipping away at the power of the Ascendancy. By 1890 the Ascendancy tradition had lost control of the Irish seats in parliament as a result of Catholic Emancipation, successive reforms of the franchise and the secret ballot; the Church of Ireland, to which the majority of them belonged, had been disestablished; agricultural tenants were guaranteed fair rent and fixed tenure; and Trinity College Dublin had voluntarily abandoned its exclusion of non-Anglicans from fellowships and scholarships.[1]

On the other hand, it was not until 1898 that local government was democratised, and the powers of patronage at county council level passed from local gentry to local nationalists. It was not until 1903 that the government actively sponsored the break-up of the larger estates among their tenants. It was not until 1908 that the state endowed the National University with an overt Catholic ethos. It was not until the constitutional crisis of 1910 demolished the House of Lords' veto that Home Rule became a realistic political possibility, and the actual transition took another decade. The Ascendancy remained dominant politically, economically and socially. Years later, W.B. Yeats was to proclaim defensively that 'we are no petty people'. In 1890 the statement would still have been superfluous.

Given the privileges of the Irish Ascendancy in education and career opportunities, it is hardly surprising that several researchers have been able to show that the majority of nineteenth-century Irish scientists had Ascendancy backgrounds. Richard Jarrell found in 1981 that between 11 and 17 of 135 individuals who taught science, worked for the government in a scientific capacity or wrote scientific articles were Catholics.[2] Bennett found that between 62 per cent and 79 per cent of the scientists in his survey were members of the Church of Ireland, and only 10.5 per cent were Catholics.[3] Gordon Herries Davies, in a less systematic tally,[4] found a slightly higher proportion of Catholic scientists, 7 out of 50. Steve Yearley agrees that 'while there appears to be a Protestant hegemony in

---

1   J.C. Beckett in *The Anglo-Irish Tradition* (1976) admits that he uses the term with reluctance, and both he and Roy Foster in their respective surveys of Irish history, *The Making of Modern Ireland 1603–1923* (1966) and *Modern Ireland, 1600–1972* (1989), tend to use 'Ascendancy' instead.

2   Richard Jarrell, 'Differential National Development and Science in the Nineteenth Century: The Problems of Quebec and Ireland', in Reingold and Rothenberg, *Scientific Colonialism* (1987).

3   J.A. Bennett, 'Science and Social Policy in Ireland in the Mid-Nineteenth Century', in Bowler and Whyte, *Science and Society in Ireland* (1997).

4   Herries Davies, 'Irish Thought in Science', in Kearney, *The Irish Mind* (1985), p. 305.

science it is not clear that it is one consistently contrived by the Protestants'.[5]

Although this last statement will be questioned below, Yearley is surely right to conclude that the nature of Ascendancy science was formed by the wider situation of the Ascendancy as a whole. Given Ireland's peripheral location and uncertain economic situation, he suggests that 'it was easy for science to become practised and valued as a *cultural* activity. Scientific attainment might be esteemed but it would be evaluated alongside other cultural accomplishments, not in terms of its utility.' This in turn limited science to the Ascendancy and also limited its potential for practical application. 'Great skill at science could readily be subculturally significant while, at the same time, there was no systematic incentive to concentrate on economically important science. From an analytical point of view, the practice of science was largely alienated from its social context.'[6]

The science that was practised in nineteenth-century Ireland was mainly carried out by members of the Ascendancy and validated by learned bodies (the Royal Dublin Society and the Royal Irish Academy) which had been created during the Ascendancy's golden era in the previous century. Part of the reason for the ultimate decline of Ascendancy science was that it was indeed culturally divisive, and could not include members of the majority Catholic community on the island apart from those few who were prepared to participate under the conditions set by Ascendancy scientists. Women were even more excluded. It is notable that no woman held a scientific teaching post in Trinity or served on the science committees of the RIA or RDS. The only women scientists among the Ascendancy tradition were astronomers with private means.

Yearley doubts that the 'Protestant hegemony in science [was] one consistently contrived by the Protestants', but that Catholics were indeed excluded from scientific participation will be demonstrated below. The results of elections to the Royal Dublin Society's Committee for Science and its Industrial Applications during the crucial period between 1915 and 1925 show that Catholic scientists were

---

5 Yearley, 'Colonial Science and Dependent Development: The Case of the Irish Experience', *Sociological Review* 37 (1989), 308–31, p. 318.
6 ibid., pp. 319–20.

much less likely to gain the status of a seat on this committee. On average they were ranked nearly four places out of twenty lower than were Irish Protestants by the electors. The Royal Irish Academy's Committee for Science was dominated by the Ascendancy until institutional reforms in 1900 enabled it to gradually become more of a partnership between the Ascendancy, state scientists and nationalists. Although Trinity College professed a non-denominational atmosphere, it retained an Ascendancy ethos and almost all of its leading figures were actively hostile to Irish nationalism.

This hostility was reinforced by the growing orthodoxy among scientists in the English-speaking world that science and religion were two mutually opposed forces, a military metaphor vigorously promoted by T.H. Huxley and his colleagues in England and by the works of A.D. White and J.W. Draper in America. Among Huxley's group, perhaps the leading challenger of theology's claims to scientific authority was John Tyndall (1820–93), originally from Leighlinbridge, Co. Carlow.[7] James Moore has shown that the depth of hostility between religion and science at this period has been exaggerated, and that almost all English scientists managed to reconcile scientific with religious beliefs.[8] But the rhetoric of rationalism, whether practised by Anglicans or agnostics, was often specifically aimed at denying the claims of the Catholic church to fix the limits of scientific enquiry or teaching. The most prominent English writer to attempt to reconcile the two, St George Jackson Mivart, ended his life rejected by both sides.[9]

Tyndall's most celebrated return to Ireland was in 1874, when as president of the British Association for the Advancement of Science he outraged both Catholic and Protestant theologians in his vehemently rationalist presidential address, given in the Ulster Hall in Belfast:

> We claim, and we shall wrest from theology the entire domain of cosmological theory. All schemes and systems which thus infringe

7   *John Tyndall: Essays on a Natural Philosopher*, ed. Brock et al. (1981).
8   James R. Moore, *The Post-Darwinian Controversies* (1979). Moore is very critical of the concept of a 'warfare between science and religion' in general but explicitly excludes Catholicism from his study (p. 11).
9   J.D. Root, 'The Final Apostasy of St George Jackson Mivart', *Catholic Historical Review* 71 (1985), p. 1.

upon the domain of science must, insofar as they do this, submit to its control, and relinquish all thought of controlling it. Acting otherwise proved disastrous in the past, and it is simply fatuous today.[10]

It is particularly revealing that, although Protestant reaction to the speech seems to have been stronger than Catholic reaction, when Tyndall wrote an apologia for the Belfast address a few months later he included a rather unconvincing *post facto* justification of his speech as an attack on the Irish Catholic hierarchy in retaliation for their supposed refusal to allow the teaching of science at the Catholic University.[11] The rationalism of the late nineteenth century provided a handy peg on which to hang a much more deep-seated prejudice for the Orangeman's son from Leighlinbridge. Needless to say, although Catholic and Protestant theologians alike objected to Tyndall's address, both used it as an opportunity to fire their own broadsides at each other.[12]

By the end of the nineteenth century the social and political dominance of the Irish Ascendancy was weakening. Part II will show how the harnessing of science to economic development was attempted not by any Ascendancy institution but by a reforming British government. But first we will investigate the fatal effect of the changing political and social context on the Irish astronomical tradition, and we will also examine the reactions of the institutions of Ascendancy science – Trinity College Dublin, the Royal Dublin Society and the Royal Irish Academy – to their new environment. The particular success of the Royal Irish Academy in overcoming challenges to its status from nationalists will be contrasted with the failure of analogous bodies in other countries to do so.

---

10 This is the wording used in the *Belfast Newsletter*'s report of the address the following day (20 August 1874). It is interesting that the pamphlet version published by Longmans later that year tones down this crucial paragraph, omitting the first sentence entirely (p. 61). The text is restored in the version published in Tyndall's *Fragments of Science, vol. II* (1879), p. 199.

11 Tyndall, *Fragments of Science, vol. II*, pp. 212–14.

12 David N. Livingstone, 'Darwin in Belfast: The Evolution Debate', in Foster and Chesney, *Nature in Ireland* (1998), pp. 387–408.

3

The Irish astronomical
tradition: rise and decline[1]

Perhaps the most striking example of the flowering of Ascendancy
science during the nineteenth century and its subsequent decline is
that of the Irish astronomical tradition. Although on the whole we
will concentrate here on the years between 1890 and 1930, astronomy in Ireland had already started to wind down by the start of this
period and an earlier starting point is therefore necessary. Nineteenth-
century astronomy is a relatively well chronicled area in the history of
Irish science. To Gordon Herries Davies' observations on the rise
and decline of Irish science in general can be added the bicentennial
histories of the Armagh and Dunsink Observatories. A number of
shorter articles about individuals and institutions have appeared in
the *Irish Astronomical Journal* and the *Journal for the History of Astronomy*.[2]

Norman McMillan has characterised the relationship between the
nineteenth-century Irish astronomers as a 'Network'.[3] Some texture

1   Much of this section has been published in *Irish Review* 17/18 (Winter 1995),
    127–41, as '"Lords of Ether and of Light": The Irish Astronomical Tradition of
    the Nineteenth Century'.
2   Most of these will be referred to in footnotes below. The best survey article is
    still Susan M.P. McKenna-Lawlor, 'Astronomy in Ireland from 1780', *Vistas in
    Astronomy* 9 (1968), 283–96, abbreviated as Susan M.P. McKenna-Lawlor,
    'Astronomy in Ireland from the Late Eighteenth to the End of the Nineteenth
    Century', in Nudds et al., *Science in Ireland 1800–1930* (1988), pp. 85–96.
3   N.D. McMillan, 'Organisation and Achievements of Irish Astronomy in the
    Nineteenth Century – Evidence for a "Network"', *Irish Astronomical Journal* 19
    (1990), 101–18.

is missing from his story, in particular that there were two periods of renewal of activity in Irish astronomy, in the 1820s and again in the 1870s. The Irish astronomers might better be described as a 'tradition' than a 'network'. Eventually the Irish astronomical tradition broke up at the start of this century. McMillan's assertion that the decline in Irish astronomy at the turn of the century was due to a split between Home Rule and Unionist astronomers does not stand up to examination of the actual political activity of the individuals concerned. It seems more likely that wider changes in Irish society and in the social structure of astronomy were responsible. John Lankford's observations, discussed below, about the changing role of amateurs within astronomy at the end of the nineteenth century are particularly relevant.

Probing the skies

Two eighteenth-century astronomical foundations survive in Ireland. One is Dunsink Observatory north of Dublin, originally constructed for Trinity College and opened in 1785, now a part of the Dublin Institute for Advanced Studies.[4] The other is the Armagh Observatory, founded in 1792 by Richard Robinson, later Lord Rokeby (1709–94), archbishop of Armagh from 1764, which is still nominally under the control of the dean and chapter of Armagh Cathedral, though closely linked to the Queen's University of Belfast.[5] It would be interesting to know what intellectual currents circulating during the time of Grattan's Parliament led to the foundation of two such institutions within a few years of each other.

Armagh and Dunsink were the only two astronomical institutions in Ireland which were attached to outside bodies – the Church of Ireland and Trinity College respectively – and this possibly accounts for their longevity. Both were fortunate in the

4   P.A. Wayman, *Dunsink Observatory, 1785-1985: A Bicentennial History* (1987), is the standard history; I have also drawn on R.S. Ball, *Great Astronomers* (1895), and *Reminiscences and Letters of Sir Robert Ball*, ed. W. Valentine Ball (1915).

5   J.A. Bennett, *Church, State and Astronomy in Ireland: 200 Years of Armagh Observatory* (1990), is an indispensable work on both Armagh Observatory and the wider Irish astronomical context. Earlier works are J.L.E. Dreyer, *A Historical Account of the Armagh Observatory* (1883), and P. Moore, *Armagh Observatory: A History, 1790-1967* (1967).

appointment of remarkable men in the 1820s: in 1823 Thomas Romney Robinson (1793–1882) became the astronomer at Armagh, and four years later the young William Rowan Hamilton (1806–65) was appointed Andrews Professor of Astronomy at Dunsink and Astronomer Royal for Ireland while still an under-graduate.[6] Robinson's greatest triumph was the publication of his star catalogue in 1859, though his most enduring innovation was the cup anemometer for measuring wind speeds, which he invented a decade earlier. Hamilton, already well known as a math-ematical prodigy when he was appointed, spent the last twenty years of his life developing the system of quaternions which he had invented in 1843. In fact his earlier contributions to dynamics have had more widespread applicability. He did not carry out a great deal of astronomical work at Dunsink, and left the routine obser-vations to his assistant and his sisters.

During the decade after the appointments of Robinson and Hamilton, two other significant astronomers emerged. William Parsons, the third Earl of Rosse (1800–67), built a three-foot reflect-ing telescope in 1839 at Birr Castle, Parsonstown (now Birr), Co. Offaly, and then went on to construct the largest telescope in the world, the 'Leviathan of Parsonstown', six feet in diameter and fifty-five feet long.[7] The great telescope's tube remains in place in the Birr Castle demesne, and the present earl is restoring it to working order as part of a science museum complex.[8] The Leviathan was built in the castle's own workshops between 1842 and 1845. Dewhirst and Hoskin comment that Rosse's set-up was almost a miniature of a twentieth-century observatory, including such crucial elements as 'a large telescope financed by a wealthy benefactor, a permanent skilled maintenance staff in optical and mechanical workshops, a small core of staff astronomers, and even

---

6   Two excellent biographies of Hamilton are R.P. Graves, *Life of Sir William Rowan Hamilton* (1882–89), and T.L. Hankins, *Sir William Rowan Hamilton* (1980). See also David Attis, 'The Social Contexts of W.R. Hamilton's Predic-tion of Conical Refraction', in Bowler and Whyte, *Science and Society in Ireland* (1997).

7   For the construction and use of the telescopes, see J.A. Bennett, 'Lord Rosse and the Giant Reflector', in Nudds et al., *Science in Ireland 1800–1930*, pp. 105–13.

8   *Irish Times*, 7 September 1996.

an occasional "guest astronomer" programme with accommodation available in the Castle'.[9]

The great telescope was built in order to settle the vexed question of the nature of nebulae, the cloudy patches of light visible here and there in the sky. The 'nebular hypothesis' was based on the ideas of William Herschel and Pierre de Laplace, as popularised by J.P. Nichol and Robert Chambers. It suggested that all these patches of light were some shining fluid, gradually condensing down to form new systems of stars and planets – some, such as the Pleiades and the Orion Nebula, clearly have stars associated with them. The idea that Nature, rather than remaining forever fixed as God had created it, was undergoing a slow evolution to a better state, had clear implications for progressive geologists, biologists and social and political theorists such as J.R. Nichol and Robert Chambers.[10]

The Leviathan of Parsonstown put the liberal adherents of the nebular hypothesis on the defensive. Rosse and Thomas Romney Robinson of Armagh convinced themselves, and the majority of other astronomers internationally, that most nebulae were composed of stars too faint and far away to distinguish individually without using the largest telescope in the world.[11] Schaffer points out that 'Robinson packaged reports from Parsonstown to do maximum damage to the nebular cosmogony.'[12] Michael Hoskin sums up the

9   David W. Dewhirst and Michael Hoskin, 'The Rosse Spirals', *Journal for the History of Astronomy* 22 (1991), 257–66, p. 257.

10 Stephen G. Brush, 'The Nebular Hypothesis and the Evolutionary Worldview', *History of Science* 25 (1987), 244–69; taken further by Simon Schaffer, 'The Nebular Hypothesis and the Science of Progress', in *History, Humanity and Evolution:Essays for John C. Greene*, ed. James R. Moore (1989), pp. 131–64.

11 Today's astronomers have shown that this is indeed true of many of the nebulae in question, which are either independent galaxies or tightly-knit clusters of stars within our own galaxy, and Rosse is justly credited as the first to realise this. Dewhirst and Hoskin, 'The Rosse Spirals', show that all of the nebulae identified by Rosse and his colleagues as spiral in form are today accepted as galaxies external to our own. However, others, including the Orion Nebula, the Ring Nebula and the Crab Nebula, which were identified by the Irish observers as composed entirely of stars, are in fact mainly gaseous and relatively near to our own solar system. The Ring Nebula and Crab Nebula are composed of the glowing gases exuded by a dying star, while the Orion Nebula is now believed to be a place of formation of new stars from the interstellar medium, just as the nebular hypothesis of the 1830s believed it to be.

12 Schaffer, 'The Nebular Hypothesis and the Science of Progress', p. 139.

Birr observations as:

> to varying degrees wish-fulfillment. Robinson badly wanted to use Irish science to discomfort English and French advocates of the nebular hypothesis, and before long the Birr reflectors were to his eyes resolving every nebula in sight; Rosse, publicly cautious but privately hoping his immense expenditure of time and money would bear fruit, was tempted along the same path; [the visiting English astronomer, Sir James] South, though looking at the same astronomical object as his two friends moments before and in the same telescope, frequently failed to see the star cluster they claimed they had just seen.[13]

Rosse's reputation has eclipsed that of his contemporary, Edward Joshua Cooper (1798–1863), who created in the early 1830s what was described at the Royal Astronomical Society in 1851 as the most richly furnished private observatory known at his home, Markree Castle, near Collooney in Co. Sligo. Under Cooper and his assistant Andrew Graham, Markree produced the standard catalogue of stars in the vicinity of the Zodiac, and also found the only asteroid, (9) Metis, ever discovered from Irish soil.[14]

The commissioning of Edward Cooper's telescope mounting for the Markree observatory marked the start of a career that would make Thomas Grubb (1800–78) and his son Howard (1844–1930)

---

13 Michael Hoskin, 'Rosse, Robinson and the Resolution of the Nebulae', *Journal for the History of Astronomy* 21 (1990), 331–44.

14 The observatory at Markree is now a depressing reminder of what can happen to scientific institutions when the tide of history turns against them. The library room was converted into a garage some decades ago, and many of the books were tipped into a hole in the floor of the meridian room next door. The meridian slit in the roof of the room is not covered, and many years of Sligo rain have found their way into the foundations, as have a number of scavengers looking for items of interest. For a description of one such expedition see N.H. Whyte, 'Digging up the Past: A Visit to Markree Castle', *Stardust* 25/4 (1992), 27–30. Some manuscript records have been rescued and are listed by Michael Hoskin, 'Archives of Dunsink and Markree Observatories', *Journal for the History of Astronomy* 13 (1982), 149–52, and Susan McKenna-Lawlor and Michael Hoskin, 'Correspondence of Markree Observatory', *Journal for the History of Astronomy* 15 (1984), 64–8. Mr John McConnell informs me that the main telescope was sold to a Jesuit seminary in Hong Kong around 1930, and that it is now in Manila.

the most successful makers of astronomical instruments in the British Isles, and perhaps the world. Burnett lists over fifty instruments made by their firm between 1868 and 1905, with customers in Australia, India, Germany, Russia and North and South America.[15] In Ireland, apart from Markree, the Grubbs supplied telescopes or mountings for Dunsink, Armagh, Queen's College Cork[16] and W.E. Wilson's private observatory in Westmeath. The pinnacle of achievement for the firm came when seven identical telescopes for seven different observatories were ordered for the *Carte du ciel* project, initiated by the Astrographic Congress held in Paris in 1887.[17]

The years around 1860 marked a dip in Irish astronomical activity with the deaths of Cooper in 1863, Hamilton in 1865 and Rosse in 1867; Robinson, though still going strong in Armagh, had been troubled by cataracts since 1859. However, the late 1860s and early 1870s saw a distinct revival. The Grubbs' regular production of telescopes began in the late 1860s. Dunsink Observatory's successive directors, Franz Brünnow (1821–91), Robert Stawell Ball (1840–1913) and Arthur A. Rambaut (1859–1923), revived an observational programme in the last three decades of the century. The two earlier private observatories were still in use. Laurence Parsons, the fourth Earl of Rosse (1840–1908), maintained the tradition of hiring astronomical assistants to use the great telescope, as did Cooper's nephew Edward Henry Cooper (1827–1902) from 1873. It is interesting that all the assistants at both Birr and Markree appointed after that date were German or Danish.[18] These later

---

15 J.E. Burnett, 'Thomas and Howard Grubb', in J.E. Burnett and A.D. Morrison-Low, *Vulgar and Mechanick: The Scientific Instrument Trade in Ireland, 1650–1921* (1989).

16 This observatory still exists and has recently been restored. There seem to be few published records of observations made from it, though I understand that it is occasionally used by UCC staff and others.

17 P.A. Wayman, 'The Grubb Astrographic Telescopes, 1887–1896', in *Mapping the Sky: Past Heritage and Future Directions (IAU Symposium 133)*, ed. S. Debarbat et al. (1988), pp. 139–42.

18 Patrick Moore, *The Astronomy of Birr Castle*, pp. 73–5, lists the Birr assistants and the dates when they worked there: W.H. Rambaut (1848–49), G.J. Stoney (1849–52), R.J. Mitchell (1852–55), S. Hunter (1860–64), R.S. Ball (1865–67), C.E. Burton (1868–69), R. Copeland (1871–74), J.L.E. Dreyer (1874–78) and O. Boedikker (1880–1916). All three of the assistants at Markree in its second period were Germans: W. Doberck (1874–82), A. Marth (1883–88) and F. Henkel (1898–1902).

assistants in Markree concentrated on meteorological work, and the Leviathan of Parsonstown fell into disuse after 1878. Otto Boedikker, the last assistant at Birr Castle, invested much time in hand-drawing a map of the Milky Way at a time when photography was making such a project obsolete, and the fourth earl preferred to use the more mobile three-foot instrument to measure the surface temperature of the moon.[19]

In the years around 1870 three more private observatories were established. John Birmingham (1816–84) began what became a seminal survey of red stars from his estate near Tuam, Co. Galway, in 1866. At Daramona House, Co. Westmeath, William E. Wilson (1851–1908) obtained his first telescope from the Grubbs in 1871, though his astronomical career peaked in the 1890s with work on photography and early photometry.[20] Finally, Wentworth Erck (1827–90) set up an observatory at Sherrington House, Co. Wicklow, in 1877 from which he published a number of scientific papers before turning his attention to politics – specifically, opposing the Land Acts – in the last years of his life.

The end of the century produced astronomers other than the professionals employed by the institutional foundations (Armagh and Dunsink) and the gentlemanly stargazers of the landed gentry. Middle-class Dublin contributed two notable observers, John Ellard Gore (*c.*1845–1910) and W.S. Monck (1839–1915), both founder members of the British Astronomical Association. Gore, who had been brought up in the neighbourhood of Markree Castle but only began observing in 1879, is recognisable as a prototype of today's amateur, dedicated to the study of variable and binary stars with only a small telescope and a pair of binoculars. Monck, a barrister who became professor of moral philosophy at Trinity, was more innovative. He experimented with the photoelectric measurement of starlight using his own selenium photo electric cells, in collaboration with W.E. Wilson of Daramona, TCD's George Francis Fitzgerald, and G.M. Minchin of the Royal Engineering

19   D. Taylor and M. McGuckian, 'The Three-Foot Telescope at Birr and Lunar Heat', in Nudds et al., *Science in Ireland 1800–1930* (1988), pp. 115–22.

20   Derek McNally and Michael Hoskin, 'William E. Wilson's Observatory at Daramona House', *Journal for the History of Astronomy* 19 (1988), 146–53.

College near Windsor.[21] Also active in the 1880s and 1890s was the Belfast-based amateur astronomer Isaac Ward (1834–1916), who discovered the 1887 supernova in the Andromeda nebula.[22]

Two Irish women from middle-class backgrounds were prominent in English astronomical circles in the last quarter of the century.[23] Lady Margaret Huggins (1848–1915), the daughter of a Dublin solicitor, became a pioneer of astronomical spectroscopy in partnership with her husband Sir William Huggins (1824–1910) at their home in Tulse Hill (using a Grubb telescope, of course); and Agnes Clerke (1842–1907), originally from Skibbereen, wrote a number of works on astronomy and cosmology including the *Popular History of Astronomy during the Nineteenth Century* for which today's historians of science are still grateful. However, professional astronomy was not yet ready to admit women as equal partners.[24] Robert Stawell Ball, at least, was prepared to regret publicly the exclusion of talented women from cis-Atlantic professional astronomy:

> It is well known that in the United States there are many more opportunities for educated women to gain useful and remunerative employment than are found in Great Britain. I was particularly struck with this when I saw American ladies employed in doing work in the astronomical observatories of their country, which on this side would be almost exclusively performed by men. The work they had to do was eminently suited to ladies; it required neatness and care, and a conscientious determination to perform it with unremitting accuracy

21 C.J. Butler and Ian Elliott, 'Biographical and Historical Notes on the Pioneers of Photometry in Ireland', in *Stellar Photometry: Current Techniques and Future Developments (IAU Colloquium 136),* ed. Butler and Elliott (1993); Ian Elliott, 'The Monck Plaque', *Irish Astronomical Journal* 18 (1987), 123–4. Fitzgerald complained of catching a cold during these experiments (letter to Oliver Lodge, 31 March 1892, in Lodge papers, University College London).

22 D.E. Beesley, 'Isaac Ward and S Andromedae', *Irish Astronomical Journal* 17 (1985), 98–102.

23 Mary T. Bruck, 'Companions in Astronomy: Margaret Lindsay Huggins and Agnes Mary Clerke', *Irish Astronomical Journal* 20 (1991), 70–7, and 'Agnes Mary Clerke, Chronicler of Astronomy', *Quarterly Journal of the Royal Astronomical Society* 35 (1994), 59–79. See also chs 3–5 of Susan McKenna-Lawlor, *Whatever Shines Should be Observed [quicquid nited notandum]* (1998).

24 See Peggy Aldrich Kidwell, 'Women Astronomers in Britain, 1780–1930', *Isis* 75 (1984), 534–46.

and attention. How successful they have been is known to all astronomers who have made themselves acquainted with the great volume of excellent astronomical research that flows from the American observatories.

I was much interested and entertained recently by reading a paper written by one of the American astronomers to whom I have referred. Two years ago a new astronomical observatory was opened at Carleton College, Northfield, Minnesota, and at the laying of the foundation-stone an appropriate – a most appropriate – address was delivered in celebration of the auspicious event. When I add that this address was given by a lady it will, I think, show that America, in respect to such performances, is much in advance of the countries on this side of the Atlantic.[25]

That Irish astronomy declined in the new century is not in dispute. Dunsink's directors after 1898 (C.J. Joly [1864–1906], E.T. Whittaker [1873–1956] and H.C. Plummer [1875–1946]) were mathematicians rather than astronomers, although a small amount of work on variable stars was carried out. In 1916, Dunsink lost its role as the centre of the Irish time service when Irish clocks were brought forward by twenty-five minutes to coincide with Greenwich Mean Time.[26] Plummer resigned in 1921, and Trinity College, due to the straitened circumstances of the time, decided not to replace him. The observatory closed until it was transferred to the state in 1947. In 1916 J.L.E. Dreyer, Robinson's successor, left to continue in greater depth the historical studies which had occupied most of his time at Armagh. Although the observatory remained open, no professionally trained astronomer was appointed to Armagh again until 1937.

Relatively close

For the country's size, there was an unusually large number of Irish astronomers in the nineteenth century, and there is no doubt that they

---

25 R.S. Ball, 'A Visit to an Observatory'; p. 97 of *In Starry Realms* (1892); the essay was originally published in *The Girl's Own Paper*.

26 Since 1968, Ireland has officially used Central European Time in summer, falling an hour behind in winter. This contrasts with the United Kingdom, which uses Greenwich Mean Time in winter and goes an hour ahead in the summer. Clocks in both countries are therefore synchronised although the legal standard time is not.

shared a common social and intellectual background. Another important factor is the network of family relationships, rather than professional patronage, which connected many Irish astronomers not just to one another but also to other scientists and Irish literary figures.

W.H. Rambaut, an assistant at both Birr and Armagh in the middle of the century, was the nephew of Thomas Romney Robinson (Armagh, 1823–82) and the uncle of Arthur A. Rambaut (Dunsink, 1892–98).[27] Mary Susanna, one of T.R. Robinson's daughters, married the physicist George Gabriel Stokes in 1857.[28] Members of the Stokes family contributed to many branches of Irish scholarship, science and medicine in the nineteenth century. Robinson's second wife, Lucy, was the daughter of Richard Lovell Edgeworth and the half-sister of the novelist Maria Edgeworth.

Robert Stawell Ball (Dunsink, 1874–92) was also a member of a remarkable scientific family. His father, another Robert Ball, had been an active member of both the Royal Irish Academy and the Dublin Zoological Society; his two younger brothers, Valentine and Charles, achieved some fame in geology and medicine respectively; and their aunt, Mary Ball, was an active naturalist into her ninth decade. R.S. Ball's first job had been as tutor to the younger children of the third Earl of Rosse, including Charles A. Parsons, who became a distinguished mechanical engineer and eventually bought the engineering company founded by Thomas Grubb. Other Rosse scientific connections included the third earl's wife, Mary, a pioneer of photography, and the microscopist Mary Ward, a cousin who was unfortunately killed in a test-drive of a steam engine (one of Charles A. Parsons' earliest designs) in the Birr Castle grounds.[29]

The physicist George Johnstone Stoney, another Birr assistant early in his career who later coined the word 'electron', was the brother of Bindon Blood Stoney, the engineer responsible for Dublin's harbour, and the father of George Gerald Stoney, who later collaborated with Charles Parsons. He was also the uncle of George

---

27 Wayman, *Dunsink Observatory*, p. 155.
28 Bennett, *Church, State and Astronomy*, p. 141.
29 For Mary Rosse and Mary Ward, see chs 1–2 of Susan McKenna-Lawlor, *Whatever Shines Should be Observed [quicquid nited notandum]* (1998). For Mary Ball, Mary Ward and two other women entomologists, see p. 232 of James M. O'Connor, 'Insects and Entomology', in Foster, *Nature in Ireland* (1998).

Francis Fitzgerald, the Trinity physicist, and his brother Maurice, who was the professor of engineering at Queen's College of Belfast.[30] The genealogical theme continues in the twentieth century – W.F.A. Ellison, director of Armagh Observatory from 1919 to 1936, was the father of Mervyn A. Ellison, senior professor of the School of Cosmic Physics at Dunsink Observatory from 1958 to 1963.

Network or tradition?

McMillan describes all these and more as members of a 'Network'. He may be making an analogy with the well-established existence of a 'Cambridge Network' in British science during the second quarter of the nineteenth century.[31] The Cambridge Network included scientists such as John Herschel, Charles Babbage, George Peacock, George Biddell Airy, William Whewell, Adam Sedgwick, John Lubbock and others. They reformed the teaching of mathematics at their own university, tried to gain control of the Royal Society, succeeded in energising the British Association for the Advancement of Science and popularised the use of the very word 'scientist'.

The Irish astronomers were active in this movement as well. Robinson and Hamilton were involved in the early years of the British Association, and are included by Jack Morrell and Arnold Thackray among the 'Gentlemen of Science' who maintained and shaped it (so is Humphrey Lloyd, later provost of Trinity College). Rosse was the first president of the Royal Society elected after the successful reform of 1847.[32] However, there is little evidence of the Irish exerting influence on one another's behalf within these British scientific bodies or even in academic appointments in Ireland and the Royal Irish Academy, of which all were members. If we are to describe them as a 'Network', some evidence of patronage would

---

30 Tom Garvin points out a possible connection also with the twentieth-century mathematician Alan Turing, whose mother's maiden name was Ethel Anna Stoney. Although the name sounds convincing, it is absent from the published Stoney genealogy.

31 The term seems to have originated with Susan Faye Cannon in *Science in Culture: The Early Victorian Period* (1978), particularly in ch. 2, 'The Cambridge Network'.

32 Morrell and Thackray, *Gentlemen of Science: Early Years of the British Association for the Advancement of Science* (1981); Marie Boas Hall, *All Scientists Now: The Royal Society in the Nineteenth Century* (1984).

help to make this concept more firm. The 'Cambridge Network' refers to a particular generation of individuals who consciously promoted their own intellectual agenda in the decades after 1830. The Irish astronomers shared only astronomy and an Ascendancy background; the analogy is a weak one. I propose instead to call them a 'tradition'.

The politics of Irish astronomy

The group of Irish astronomers who became active in the 1820s – Thomas Romney Robinson, Edward Cooper, the third Earl of Rosse and William Rowan Hamilton – had a common Ascendancy background. However, they did not share identical political outlooks. Cooper and Rosse were both MPs during the 1830s, but sat on opposite sides of the House – Rosse was a Whig, but Cooper and Robinson were Tories, as was Hamilton when he turned his mind to such mundane matters.[33] While Rosse was relatively cautious about claiming new discoveries made with the Leviathan, Robinson could barely be restrained from proclaiming the great Birr telescope's success at rendering visible the component stars of the nebulae as a triumph of the Irish national genius against the (liberal) nebular hypothesis.[34]

McMillan portrays the decline of the 'Network' as the result of the breakdown of a political consensus (whose existence is doubtful) among Ascendancy scientists into a 'bitter internecine struggle' with 'a group of "Home Rulers" in the Royal Irish Academy', including the Trinity Fellows, J.A. Galbraith and Samuel Haughton, becoming estranged from their colleagues in the Unionist camp led by George Johnstone Stoney. He states that, under pressure from Stoney, the fourth Earl of Rosse 'abandoned his family's long-held nationalist politics and moved into the Unionist camp'.[35] He may be drawing from Webb and McDowell, who say of the fourth earl around 1900

---

33 For Hamilton's politics see ch. 15, 'Reform and Religious Turmoil', of Hankins' biography.
34 Michael Hoskin, 'Rosse, Robinson and the Resolution of the Nebulae', *Journal for the History of Astronomy* 21 (1990), 331–44; see also Michael Hoskin, 'The First Drawing of a Spiral Nebula', *Journal for the History of Astronomy* 13 (1982), 97–101.
35 McMillan, *op. cit*., p. 113.

that 'the events of the past twenty years had made him desert the liberal traditions of his family for a pessimistic and disillusioned Toryism'.[36]

But the Parsons family's aristocratic Whiggery cannot possibly be described as 'nationalist'. As a young MP, the third earl had supported Catholic Emancipation and the Maynooth grant, and was a late convert to Reform. But he also wrote a pamphlet and spoke in parliament opposing the repeal of the Corn Laws, and there is no evidence to suggest that he had any sympathy for O'Connell's Repeal movement or for Young Ireland.[37] Another pamphlet, written in the midst of the Famine, condemns new-fangled political economists equally with the 'profoundly ignorant' opponents of Newtonian philosophy.[38] Certainly, by the time he had graduated to the House of Lords as an Irish representative peer he was to be found supporting the Irish policies of Peel's Tory administration, although he favoured closer government supervision rather than local control of the state's faltering efforts at famine relief.[39] As for the fourth earl, Webb and McDowell imply that his political conversion was a result of disenchantment with Balfourian Unionism and the weakening of the Ascendancy in the 1880s and 1890s, rather than moral pressure from Stoney in the 1870s. Rosse's reaction to 'reform' was not unusual – the impact of 'constructive Unionism' on Unionist voters in Dublin was so negative that both their parliamentary seats were lost in the 1900 general election.[40]

In any case, the timing of events does not fit McMillan's thesis. Irish astronomy was flourishing during the last three decades of the

---

36 McDowell and Webb, *Trinity College Dublin, 1592–1952: An Academic History* (1982), p. 410.

37 For the third earl's political career, see *Dictionary of National Biography*; see also Hansard, *Parliamentary Debates, 2nd series,* 21, 1692 (2nd June 1829), for a speech against parliamentary reform; see ibid., 13, 237 (28 April 1825), for a speech opposing repeal of the Corn Laws; see ibid., *3rd series,* 18, 566–73 (11 June 1833), for a clash with Daniel O'Connell.

38 Rosse, *Letters on the State of Ireland* (1847), p. 10.

39 See Hansard, *Parliamentary Debates, 3rd series,* 81, 230 (9 June 1845), and especially 81, 581–4 (16 June 1845); on the Famine, see ibid., 91, 484–6 (26 March 1847), 92, 440 and 461 (6 May 1847), 94, 234–5 (13 July 1847).

40 Alvin Jackson, 'The Failure of Unionism in Dublin, 1900', *Irish Historical Studies* 26 (1989), 377–95.

century. The Home Rule debate was happening at exactly the same time, so it can hardly therefore have killed off the astronomical tradition. (It is interesting that the earlier peak of activity after 1825 coincides with the political rise of Daniel O'Connell.) If anything, the Home Rule issue gave the astronomers and other scientists who had Protestant backgrounds a political unity which they had previously lacked. Apart from the fourth Earl of Rosse, one could name Lord Kelvin, Sir George Stokes and Sir Robert Ball as political liberals who moved, with Joseph Chamberlain, firmly into the Unionist camp after 1885. Tom Dunne has pointed out that this was a common reaction among British liberal intellectuals.[41]

The decline of Irish astronomy

The reasons for the death of Irish astronomy lie not in politics but in economics and in the changing requirements of the scientific enterprise. The decline at the end of the century began when the ownership of land began to be transferred from landlord to tenant. The Rosses and Coopers could no longer support an astronomical assistant, and the Birminghams, Wilsons and Ercks found that their income too was dwindling – Erck indeed spent his final years in ultimately futile opposition to compulsory land-purchase.

But even more significant was the change in the very nature of the way astronomy was practised worldwide. Astronomy in Ireland had never become very professionalised. Robinson was the only astronomer at either Armagh or Dunsink ever to assume a position of leadership within Irish astronomy. By the turn of the century, the future for astronomy was in astrophysics. Amateurs such as the Hugginses and Monck could and did strike out in new directions, but they were now in a position where their work had to be validated by the professionals who were defining the science. The less affluent amateurs such as Gore and Ward could not afford to stay at the cutting edge. As John Lankford puts it, in his study of a number of amateur astronomers from England and America in the late nineteenth and early twentieth centuries, 'neither wealth nor leisure

---

41 Tom Dunne, '*La trahison des clercs:* British Intellectuals and the First Home-Rule Crisis', *Irish Historical Studies* 23 (1982), 134–73.

could take the place of specialized training to provide access to increasingly sophisticated technologies'.[42]

Perhaps, if the two permanent astronomical institutions in Ireland had modernised and expanded in time, they could have maintained a native astronomical tradition. In fact Armagh Observatory's resources were so severely reduced by the disestablishment of the Church of Ireland in 1869 that Robinson was moved to complain bitterly shortly before his death in 1882 of 'the indifference of Mr Gladstone to anything which does not forward the interests of Revolution'.[43] The observatory did receive a once-off government grant of £2,000 the following year, but it had to wait almost another half-century before receiving any regular state income and was never in a position to consider making the capital investments necessary to stay at the forefront of astronomy. Dunsink Observatory, as part of Trinity College, was not as badly affected by the reforms of the later nineteenth century. However, only two (Brünnow and Rambaut) of its six directors after 1865 were primarily interested in astronomy rather than mathematics. The office of Astronomer Royal for Ireland had become identified more with mathematical excellence than with observational astronomy under Hamilton, and Ball, Joly and Whittaker maintained that side of the tradition. Ireland as a whole only began to catch up with the rest of the astronomical world under the leadership of Armagh's Eric Lindsay in the late 1930s and with the foundation of the Dublin Institute for Advanced Studies in 1940.

Irish astronomy as Ascendancy culture

The pursuit of knowledge of the natural world as primarily a cultural activity, rather than a means to social and economic development, meant that it was vulnerable to social change, and the construction and maintenance of expensive telescopes by the elite was particularly so. Astronomy as a cultural activity was not divorced from either wider culture or political discourse, as demonstrated by Rosse and Robinson's use of the Leviathan of Birr to 'disprove' the nebular hypothesis. The main forum for the validation

---

42 John Lankford, 'Amateurs and Astrophysics: A Neglected Aspect in the Development of a Scientific Specialty', *Social Studies of Science* 11 (1981), 275–303.

43 Quoted in Bennett, *Church, State and Astronomy*, p. 152.

of astronomical work was the Royal Irish Academy, which was also engaged in the literary and antiquarian work that would lead to the Gaelic revival. To understand the Irish astronomical tradition, we must not only consider its location in worldwide astronomy but also its place in Irish culture. Perhaps writers on nineteenth-century Irish culture might also profitably include the contribution of the astronomers who so firmly located themselves within it.

4

Trinity College:
fighting off reform

Trinity College, the sole constituent college of the University of Dublin, is the oldest Irish university, founded by Elizabeth I in 1592.[1] Long identified as the intellectual territory of the Ascendancy, its position had become increasingly precarious as the question of university education became identified as a sectarian issue in nineteenth-century Irish politics. A fuller account of the Irish University Question will be given in part III, but the various efforts, both internal and external, to change Trinity College will be outlined here. The activities of Trinity's scientists, both in attracting resources and in terms of their own research activities, will also be reviewed. In particular, the careers of George Francis Fitzgerald and John Joly will be outlined. Finally the changes brought about in Trinity by the advent of the Irish Free State will be described.

Legislation in the 1790s had allowed Catholics to take degrees at Trinity College – they had been excluded altogether since 1637 – but

---

1 Trinity College has been fortunate enough to generate a number of institutional histories. The latest is R.B. McDowell and D.A. Webb, *Trinity College Dublin, 1592–1952: An Academic History* (1982). K.C. Bailey, *A History of Trinity College Dublin, 1892–1945* (1947), was published at a point when the college was reopening negotiations with the Irish government after some twenty-five years. Wesley Cocker, 'A History of the University Chemical Laboratory, Trinity College Dublin: 1711–1946', *Hermathena* 124 (1978), 58–76, concentrates on the earlier period.

they were not eligible for scholarships or fellowships. In 1873 Gladstone's government proposed that the University of Dublin be expanded to include all the higher education colleges in Ireland – Trinity, the Queen's Colleges, the Catholic University, Magee College in Derry, and any others that met the plan's criteria. His bill was defeated by three votes in the House of Commons, and the government fell as a result. In the immediate aftermath, Henry Fawcett, the blind postmaster-general, managed to persuade parliament to abolish all religious tests in Trinity College. Fawcett faced opposition from Irish Catholics, who felt that it merely allowed them an inferior place in an overwhelmingly Anglican institution, and also from many in England, and many Irish Protestants, who saw it as a dangerous concession so soon after the disestablishment of the Church of Ireland in 1869. However, he had the crucial support of Trinity College itself, which recognised that disestablishment must bring an end to its previous exclusive Anglicanism. As has been pointed out above, Ascendancy learning was now eager to portray itself as non-sectarian and free from religious bias.

By the end of the century, Catholic demands still had not been met. The Robertson Commission of 1901–3 recommended that the Royal University expand to include a new Catholic college in Dublin, and that the Galway and Cork colleges be made more acceptable to Catholics. Trinity College however had been explicitly excluded from its terms of reference, and several of the commissioners made it clear that they believed that the new Catholic college should if possible be included in the University of Dublin, alongside TCD, rather than in the Royal University, alongside the Queen's Colleges. In an attempt to avert this, the board of Trinity proposed to provide a Catholic chapel, chaplains and religious instruction classes. The hierarchy on the other hand wanted a Catholic university or an explicitly Catholic college within a university, and rejected the Trinity plans as insufficient to meet their demands.[2]

---

2   T.W. Moody, 'The Irish University Question of the Nineteenth Century', *History* 43 (1958), 90–109, and also Fergal McGrath, 'The University Question' (1971) pp. 85–142.

The Bryce plan of 1907

The Liberal government which came to power at the end of 1905 set up the Fry Commission to report on the position of Trinity itself in the structure of Irish higher education. The commission reported at the start of 1907, and recommended that a Catholic college be established in Dublin. But they were split between the options of either linking it with the Royal University or repeating Gladstone's proposal that all Ireland's colleges be united under the umbrella of the University of Dublin. Bryce, the chief secretary for Ireland, immediately adopted a version of the second proposal as government policy. Trinity College, the new college (including the laboratory facilities of the state-controlled Royal College of Science) and the Queen's Colleges of Cork and Belfast were to be incorporated into the University of Dublin. The Galway college was to be affiliated to the new body.

Bryce's plan was endorsed by the bishops and by nationalists more generally. He himself left Ireland as the new British ambassador to the United States a few days after proposing it, leaving his successor, Augustine Birrell, with a policy which was bitterly opposed by Trinity College and received only lukewarm support from the other Irish colleges. Particularly contentious was Bryce's proposal to divide responsibility for teaching different subjects between the (denominational) colleges and the (non-denominational) university in the new institution, so that 'the University would provide teaching in advanced subjects which are non-controversial subjects, such as mathematics. You cannot have Protestant mathematics and Catholic mathematics.'[3]

Trinity objected that the Catholic hierarchy would effectively be allowed to determine which subjects were controversial and which were not, and quoted evidence given by Catholic clerics to the two recent Royal Commissions to demonstrate that no subject could be considered safe from priestly meddling:

> It is interesting to compare this statement of Mr Bryce's [quoted above] with the following extract from a paper by the Reverend

---

3   Bryce's speech of 25 January 1907, reported in both *Irish Times* and *Freeman's Journal* the next day

Professor D. Coghland, DD, of Maynooth, who writes (Appendix to Final Report of Fry Commission, p. 415): 'There *can* be grave danger from non-Catholic teachers and fellow-students, even in purely secular or scientific classes. A teacher *could*, for example, make a covert hostile allusion to the principle of authority in the church even when teaching mathematics, by remarking significantly that mathematical conclusions are not received on authority, that scientific work and authority are mutually incompatible.'[4]

E.J. Gwynn, a Celtic scholar and future provost, and E.P. Culverwell, once a mathematician but now the professor of education, led Trinity's campaign. It ran through most of February and March 1907. Public meetings of Trinity graduates were held, politicians in London were lobbied, and the voices of other universities were raised in defence of Trinity. Significantly, the first of these was the University of Cambridge, where a meeting of Trinity supporters was held on 2 March. Sir Robert Ball, the former director of Dunsink Observatory who had moved to Cambridge in 1892, chaired the event and the invited speaker from Trinity was John Joly, the professor of geology. A number of mathematicians and physicists from Irish Ascendancy backgrounds had migrated to Cambridge over the years, including Ball himself, George Gabriel Stokes, and, slightly later, Joseph Larmor.

The case for maintaining the separate status of the University of Dublin was based on its self-professedly non-denominational character. It was argued that Catholic students who wished to attend it did so on an equal basis with Protestants, and the fact that few did so was purely down to the attitude of the Catholic bishops. Trinity saw its non-sectarianism as linked with the true spirit of scientific enquiry:

> The intellectual liberty which we uphold cannot be reconciled with that educational theory which the heads of the Roman Catholic Church have defined as founded on the principle of authority. If University teaching is to be shared by Colleges which hold conflicting views, there must be constant occasions of strife and bitterness,

---

4   [E.P. Culverwell], *Mr Bryce's Speech on the Proposed Reconstruction of the University of Dublin: Annotated Edition issued by the Dublin University Defence Committee* (1907), p. 27.

which can only end when one or other of the contending parties loses its distinctive character. It is contrary to our best traditions that the boundaries of science should be fixed, directly or indirectly, by ecclesiastical authority, or the impulse of speculation arrested by clerical intervention. Between two or more Colleges so sharply divided by principle and tradition there can be no real union.[5]

The new chief secretary for Ireland, Augustine Birrell, soon let it be known that Bryce's plan was to be dropped. His career in Ireland, which was to end with the humiliation of the Easter Rising in 1916, began auspiciously with a solution to the University Question which satisfied everyone except the Ulster Unionists. Trinity College was to be left untouched; Queen's College Belfast was to become a university in its own right; the Queen's Colleges in Cork and Galway, renamed University Colleges, were to be united with University College Dublin in a new National University of Ireland which would be Catholic in atmosphere. Maynooth was to affiliate with the National University, and Magee College with Queen's (though in the event it decided to make a formal link with Trinity instead). This solution was a compromise between the Catholic hierarchy, which had dropped its earlier demands for close control of a Catholic university, and the government, which had weakened on the principle of not endowing denominational education. The two universities set up in 1908 have survived to the present day, as has Trinity.

Fitzgerald and Joly

Apart from the question of its relationship to other Irish higher education institutions, Trinity faced another problem around 1900. Throughout the nineteenth century there had been no appreciable growth in the college's income, and the college was unable to keep pace with the growing cost of adequately equipping its scientific laboratories. The two leading activists in attempts to raise money from various sources for the further endowment of the college's scientific

---

5   Dublin University Defence Committee, 'Trinity College, Dublin, and the Proposed University Legislation for Ireland', par. 6; quoted from the *Freeman's Journal*, 11 March 1907.

work were George Francis Fitzgerald (1851–1901) and John Joly (1858–1933), respectively the College's professors of natural philosophy (i.e. physics) and of geology.[6]

Fitzgerald's efforts to improve the college's scientific facilities were halted by his early death in 1901. John Joly took on the role of chief fund-raiser for the Science Schools Committee. He and Fitzgerald had shared a mistrust of government-funded research, and perhaps missed an opportunity to exploit the sympathetic government of Lord Salisbury and Arthur Balfour for this very reason.[7] Joly had turned down the chance of a post in the Kew Observatory in 1893. Fitzgerald wrote to Lodge that if Joly had taken the Kew job, he 'would not be allowed to use "Kew Gas, Kew Water and Kew appliance" for his own work. If that is the spirit of the place it would not suit him. It is the spirit of all official places hence the objection to National Laboratories, even Home Rule wont [*sic*] drive our original man into one.'[8]

Rather than beg from the state, the college appealed to its graduates through Joly and the fourth Earl of Rosse. Only 235 responded to the appeal, compared with 1,700 who had contributed to the Tercentenary Fund to provide accommodation for student societies a few years before. K.C. Bailey, himself a chemistry professor, describes this as 'a curious commentary on the relative unimportance of scientific development in the eyes of the graduates of 1903'.[9] However, £20,000 was raised for further scientific research, a sum more than matched by the Earl of Iveagh's donation of £24,450, the cost of building new physics and botany laboratories. The question of increasing the science facilities was not raised again until after the 1914–18 war.[10]

George Francis Fitzgerald was the son of a Church of Ireland bishop, and the nephew of the physicist George Johnstone Stoney.

---

6 For Fitzgerald see esp. McDowell and Webb, *Trinity College Dublin*, pp. 305–7; for Joly, ibid., pp. 408–10, also T. de Vere White, *The Story of the Royal Dublin Society* (1955), pp. 202–5, and J.R. Nudds, 'John Joly: Brilliant Polymath', in Nudds et al., *Science in Ireland 1800–1930* (1988), pp. 163–78.

7 See D.S. Cardwell, *The Organisation of Science in England* (1972), pp. 202–3, for the increase in funding for scientific purposes gained by English universities between 1897 and 1914.

8 Fitzgerald to Lodge, 12 April 1893 (Lodge papers, University College London).

9 Bailey, *Trinity College Dublin*, p. 61.

10 McDowell and Webb, *Trinity College Dublin*, pp. 405–11.

His brother Maurice became the first professor of engineering of the then Queen's College Belfast and his wife, Harriet, was the daughter of his predecessor as professor of natural philosophy, John Hewitt Jellett, who became the provost of Trinity College in 1881. Fitzgerald's main interest was in Maxwellian electromagnetism. He is best remembered for his explanation of the unsuccessful result of the Michelson-Morley experiment to determine the relative velocity of the earth and the ether. Fitzgerald proposed that moving objects contract in the direction of their motion and that moving clocks run slow, and wrote what became two of the three essential equations of Einstein's Special Relativity.[11]

Fitzgerald was a man of many enthusiasms. He is affectionately remembered in his college for his attempts to build a flying machine (actually a glider, launched by the pulling power of several students) and for his activism in the Science Schools Committee. Outside the college he was also a promoter of science teaching, and was one of the movers behind the foundation of the Kevin Street Technical School in Dublin. His political activities seem to be less celebrated. Fortunately, they are illuminated in his letters to his friend Oliver Lodge, which survive in University College London. The two physicists were close, and addressed each other as Φ and Λ (one of Fitzgerald's letters ends '. . . with best regards to Mrs Λ and the λs').[12]

Like most of the staff at Trinity, Fitzgerald opposed Home Rule for Ireland. McDowell and Webb do not single him out as especially active in politics during this period, but he wrote to Lodge during the time of the second Home Rule bill (1892–93) that 'with this Home Rule looming in the near future as possible I cannot afford to wast[e] more money than I can help as I shall almost certainly have to leave Ireland if it comes on'.[13] The hostility of Fitzgerald and his Trinity colleagues to Home Rule can be rationalised in terms of self-

---

11　Bruce J. Hunt, 'The Origins of the FitzGerald Contraction', *BJHS* 21 (1988), 67–76; Stephen G. Brush, 'Note on the History of the FitzGerald-Lorentz Contraction', *Isis* 58 (1967), 230–2; Alfred M. Bork, 'The "FitzGerald" Contraction', *Isis* 57 (1966), 199–207.

12　Fitzgerald to Lodge, 23 March 1894 (Lodge papers, University College London).

13　Fitzgerald to Lodge, 11 July 1892 (Lodge papers, University College London).

interest. They saw no attraction whatever in an Irish government certain to be dominated by Irish Catholics, members of a church which was hostile to the University of Dublin, and committed to removing the remaining privileges of the Ascendancy class. In Fitzgerald's case at least, this was backed up by a hostility to Catholics generally:

> I was most interested to hear a gentleman in the train who came to meet a lady, both in [the] First Class carriage, both R.C.s say to her that the only reason he objected to a place she had been stopping at near Richmond on the Thames was because one of the roads to it passed a graveyard and that he would not for worlds pass it at night. The young lady said that she was more afraid of the living than the dead but the old gentleman demurred to that sentiment on his part. When this class is so superstitious what can you say about the ignorant? They universally are more afraid of and expect more help from dead saints than anything else. It really prevents them from rational activity almost as much as the fetish worship of an African savage.[14]

Such sentiments cast new light on Fitzgerald's enthusiasm for educating the working men of Dublin as a possible means of bringing them to enlightenment, i.e. away from the fetish-worship of Catholicism, through evening classes.

Like Fitzgerald, John Joly was the son of a Church of Ireland cleric, though his mother, an Austrian countess, added an exotic dash to his background. He too had other scientific relatives – his cousin, Charles Jasper Joly, was the director of Dunsink Observatory from 1898 until his early death in 1906. Unlike Fitzgerald, he never married, and it seems probable that he had a number of homosexual relationships with his students and close colleagues.[15] After graduating in 1882, Joly held various teaching posts in Trinity before being appointed professor of geology in 1897. Although he became known as a geologist, he might be better described as an early geophysicist. In 1899 he estimated the age of the earth at 80–90

---

14 Fitzgerald to Lodge, 21 December 1896 (Lodge papers, University College London).

15 This is supported by oral tradition at Trinity College, and by inference from much of Joly's (mostly incoming) correspondence in the college manuscripts collection.

million years, basing his calculation on the rate of increase of the sodium content of the oceans. Later in his career, his work on the formation of pleochroic haloes around radioactive inclusions in minerals such as biotite paved the way for what has now become the standard method of measuring the ages of rock formations.

Joly's interests extended into other areas of physics. In collaboration with his close friend H.H. Dixon, the professor of botany, he origi- nated the idea that the ascent of sap in plants is due to the cohesive strength of water. Early in his career, he invented the steam calorime- ter for measuring the specific heats of various materials, and in 1895 he produced a new method of colour photography. During the 1914–18 war he submitted numerous plans for inventions to the British admiralty, few if any of which seem to have been utilised. More successful was his plan to use the radioactive gas then called Emana- tion (and now called radon) for therapeutic purposes from 1912.

Unlike Fitzgerald, Joly lived long enough to see the threat of Home Rule swamped by the violence of the 1916–23 period. He wrote a graphic and moving description of the defence of Trinity College during the 1916 Rising.[16] Terence de Vere White, who knew Joly towards the end of his life, describes him as 'blind and deaf to nation- alist sentiment which he called lawlessness and sedition'.[17] His correspondents shared this outlook: one former pupil working in Tasmania wrote in 1922 that 'TCD must be the only area within the City of Dublin, which has not gone stark, staring mad during the last 6 years or so.'[18] Among Joly's souvenirs are his membership card for the Irish Loyal and Patriotic Union (founded 1885, reorganised and renamed the Irish Unionist Alliance in 1891) and some charred registered mail certificates recovered by Joly from the ruins of the General Post Office in May 1916.

The new dispensation

By the end of the war there had been a significant shift in the posi- tion of the Southern Irish Unionist establishment of which Trinity

---

16 'In Trinity College During the Sinn Féin Rebellion, by One of the Garrison [John Joly]', *Blackwood's Magazine* 200 (July 1916), 101–25.

17 White, *RDS*, p. 204.

18 TCD MSS 2312/257: W.B. McCabe to Joly, 20 February 1922.

was a central part. The Irish Convention of 1917, comprising 100 representatives of various political parties, churches and other interests, was set up as an attempt to resolve the Irish Question after the 1916 Rising. It met in the Provost's House at Trinity College, chaired by the veteran 'constructive Unionist' Sir Horace Plunkett (whose earlier career will be the subject of discussion in a later chapter). The convention foundered on the question of the financial autonomy of an Irish government, with Ulster Unionists willing to concede no ground at all and nationalists split between pushing for complete Irish control and accepting a more moderate proposal from Southern Unionists. With the rise of Sinn Féin as a political party, due to its being credited by the authorities with responsibility for the rising, it was clear to most former unionists outside Ulster that their future could only be assured by some measure of accommodation with the forces of separatism.

For the new provost of Trinity, the former archbishop of Dublin John Henry Bernard, an even more pressing worry than the broader political outlook was the college's dire financial situation, brought on by wartime inflation and a dearth of students. Along with Oxford and Cambridge, Trinity turned for the first time in its recent history to the state. A Royal Commission was appointed, chaired by the geologist Sir Archibald Geikie, who had been a strong supporter of the university's 1907 campaign, and with John Joly as another of its members. The commission's report, considerably briefer than its predecessors of 1907 and 1903, was very sympathetic. Higher salaries were recommended all round; large capital grants should be made to the departments of chemistry, engineering, physiology and pathology; additional lecturing posts should be created, especially in those four departments and in zoology, botany, physics and geology. Altogether a capital grant of £113,000 and an annual grant of £49,000 would be needed (full details are given in Appendix 3).[19]

Under the Union as it had been, a Westminster government would certainly have cut down on the total sums of money but generally treated the commission's report sympathetically, as happened with the contemporaneous Royal Commissions on Oxford and

---

19 *Report of the Royal Commission on the University of Dublin*, 1920 (Cmd 1078: 1920 XIII 1189).

Cambridge. Indeed, the British Treasury made a series of non-recurrent grants totalling £56,000 in 1919–23. But the changed political situation in Ireland made anything approaching a fuller implementation of the Royal Commission's report impossible.

In the general election at the end of 1918, Sinn Féin won 73 of the 102 Irish seats. These elected MPs refused to sit in Westminster and constituted themselves as Dáil Éireann, an Irish parliament. A campaign of civil disobedience followed, accompanied by terrorism from both nationalists and British armed forces. This left the government committed to implementing Home Rule while attempting to defeat the militant nationalists and to come to terms with the refusal of Ulster Unionists to entertain an all-Ireland government. The 1920 Government of Ireland Act partitioned the island into two autonomous areas, Northern and Southern Ireland. The Act committed the Southern Irish parliament to an annual grant to Trinity College of £30,000.

However, this Southern Ireland parliament never functioned. When the elections for its House of Commons were held in 1921, all 124 members were elected unopposed. Apart from Trinity College's four representatives, all were supporters of Sinn Féin who refused to operate the system. The four Trinity College representatives, and a slightly larger number of senators, met twice in June 1921 as a stunted parliament. By this stage, however, the government was engaged in the negotiations with Sinn Féin that resulted in the end of the war in July and the Anglo-Irish treaty of December 1921 which gave Southern Ireland dominion status as the Irish Free State.

Despite verbal reassurances from Lloyd George and Arthur Griffith, the treaty contained no particular safeguards for Trinity College. As Webb and McDowell point out, 'it is very doubtful if the College would have benefited in the long run from being the beneficiary of a grant unwillingly paid by the Irish government as part of a bargain for independence'.[20] Provost Bernard was occupied for most of 1922 in lobbying on behalf of Southern Unionists more generally, and it

---

20 Webb and McDowell, *Trinity College Dublin*, p. 427. My interpretation of the chronology of Trinity's negotiations with the Irish Free State government differs from theirs. Robert H. Murray, *Archbishop Bernard: Professor, Prelate and Provost* (1931), is a rather partisan and incomplete contemporary account.

was only late in the year that he approached the new government on behalf of the college in particular. His first reply from Cosgrave, who was both the head of government and the minister for finance,[21] was 'that the Provisional Government are not bound by anything in Section 64 of the Government of Ireland Act, 1920'.[22]

The college tried again in March and April 1923.[23] At first it drew an astonishingly frosty response from Joseph Brennan of the Department of Finance, demanding to know full details of the college's expenditure on salaries, food, accommodation, student prizes, sporting activities, its income from student fees in comparison with Oxbridge, and the availability of its accounts.[24] The board replied quite mildly that 'full information on all the points indicated was submitted to the Royal Commission which investigated the needs of Trinity College in the year 1920'.[25] Brennan responded that UCD was now to get preferential treatment, due to its historic underfunding. He did offer a one-off grant of £5,000 but in terms which were quite unacceptable to the board.[26]

At this point the board took the surprising line of consulting the governor-general of the Irish Free State, the veteran Nationalist politician T.M. Healy. Earlier in the year, the college had presented a loyal address to Healy[27] and clearly expected a sympathetic hearing from him. The governor-general had a secret meeting with the provost on the evening of 22 June,[28] and perhaps at his suggestion the board decided to threaten the government

---

21 Cosgrave took on both these offices after the deaths of Arthur Griffith and Michael Collins in August 1922; he relinquished the ministry of finance to Ernest Blythe in August 1923.

22 Cosgrave to Bernard, 22 November 1922 (TCD MSS 2388–93/379).

23 Webb and McDowell, *Trinity College Dublin*, pp. 427–8, have the negotiations starting in May; but apart from the brief exchange of November 1922, Bernard contacted the government again on 14 March (TCD MSS 2388–93/400) and 11 April (TCD MSS 2388–93/405) 1923.

24 George Brennan (which must be a mistake for Joseph Brennan) to W. Thrift, TD for Trinity, 31 May 1923 (TCD MSS 2388–93/420).

25 [Board] to Cosgrave, [12th] June 1923 (TCD MSS 2388–93/431).

26 Brennan to 'A Dhuine Uasail' (the Provost), [*c.* 18] June 1923 (TCD MSS 2388–93/432).

27 *Irish Times*, 7 February 1923.

28 Healy to Bernard, 22 June 1923 (TCD MSS 2388–93/423).

with publicity in a letter sent the next day:

> The Board of Trinity College have received a letter from the Ministry
> of Finance, dated June 1923, and have been greatly surprised by its
> tenor. It would be impossible for any academic authority to accede to
> the terms of that letter and they think it desirable that the matter
> should be further considered by the Government, before publicity is
> given to the correspondence.
>
> We understand that Mr Thrift has already placed before you the
> situation, as it presents itself to the College authorities, and also has
> explained the urgent financial needs of the University.[29]

By lunchtime that day, the governor-general was able to tell the
provost that 'the document of which you sent me [a] copy will in
effect be withdrawn, as I understand a new letter is in course of
gestation'.[30] Brennan's official response, on behalf of the ministry of
finance, was not exactly graceful:

> I am directed by the Minister of Finance to state that he has had an
> opportunity of discussing this subject further with the representatives
> of the University in the Dail and, in view of the further facts placed
> before him, is prepared to modify the proposals made in my letter of
> the 18th in the manner explained below.[31]

The college was to get the £5,000 grant with almost no strings
attached, and a financial muddle left over from the 1903 Land Act
was to be resolved in the college's favour, to the tune of £76,000 in
investment capital and a £3,000 annual grant. The college did not
approach the Irish government again until 1945. Healy's personal
intervention passed unnoticed by historians.[32]

The most obvious scientific casualty of the new financial regime
was Trinity College's observatory at Dunsink. Henry C. Plummer,
who had been its director since 1912, resigned in 1921 to take up a

---

29 [Board] to [Cosgrave], 23 June 1923 (TCD MSS 2388–93/425).

30 Healy to Bernard, 12.30 p.m. 23 June 1923 (TCD MSS 2388–93/424).

31 Brennan to 'A Dhuine Uasail', 27 June 1923.

32 Webb and McDowell seem to have concentrated on other sources than
    Bernard's papers for this episode (*TCD,* p. 428 and notes, p. 553); they would
    certainly have commented on Healy's involvement had they known of it.

post at the Woolwich military academy which was better paid. The decision of the board not to appoint a replacement may have originated with the observatory assistant, Charles Martin. In late September he wrote to Joly:

> In a conversation I had with you in reference to the Observatory, you gave me some good advice, you also told me to please myself as to what steps I took in the matter, so I am writing to let you know, that I informed the Provost as to the state of the Observatory. The place is in such a bad state, that something must be done at once in the way of repairs, and I thought, that if the post here could remain vacant for some time, the money saved would help to put the Observatory in a much better condition, again by waiting until the country was in a more peaceful state, the College would have their choice of good me[n]. The Provost was exceedingly Kind to me, and suggested that I should let him have the sum required for repairs.[33]

The post did indeed remain vacant for 'some time'. Dunsink Observatory remained closed until 1950, when it reopened as part of de Valera's Institute for Advanced Studies. The Andrews Professorship was revived as the title of a personal chair for the then director of Dunsink in the late 1970s, but there has been no Astronomer Royal for Ireland since 1921.

---

33 Martin to Joly, 23 September 1921; Trinity College Dublin MS 4796/124.

5

Royal Dublin Society:
compromise and change

Another institution with roots firmly based in the old order was the Royal Dublin Society, founded in 1731 by a group of activists including several chemists and the physician and writer Thomas Molyneux, who had been active in the Dublin Philosophical Society which had flourished in the 1680s.[1] The purpose of the society was declared to be the improvement of 'Husbandry, Manufactures and other useful arts and sciences'. During the next 150 years, it introduced scientific initiatives such as the Botanic Gardens at Glasnevin and the Museum of Irish Industry, which evolved into the Royal College of Science, and its collections formed the core of what became the National Museum. In the course of the nineteenth century these were taken over by the government, and their story is told in part II.

After it was relieved of responsibility for what became the Science and Art Institutions, the RDS continued to be active in the field of science as well as in its programme of annual agricultural shows. The

---

1   Several books have chronicled the RDS, including H.F. Berry, *A History of the Royal Dublin Society* (1915). I have mainly drawn on the last two chapters, pp. 295–327, by R.J. Moss; T. de Vere White, *The Story of the Royal Dublin Society* (1955); and most recently *RDS: The Royal Dublin Society, 1731–1981* (1981), edited by J. Meenan and D. Clarke. *Nostri Plena Laboris* (1987), edited by Charles Mollan et al., is an author index to the RDS scientific journals from 1800 to 1985. The society's records are in very good order in its library at Ballsbridge.

activities of the society's Committee for Science and its Industrial Applications between 1889 and 1920 are surveyed here. These largely took place under the guidance of John Joly, whose long association with the RDS culminated with his term of office as president during its bicentennial celebrations. The role of the society as a gatekeeping institution which prevented Catholic scientists from gaining the same access to resources and to social status as their Irish Protestant and English colleagues will be examined. So will the reasons for the decline of the society's scientific activities in the period from 1920 when, after a faltering start, the RDS membership and its promotion of agriculture entered a period of unprecedented growth.

The year 1890 saw a settlement of a number of issues which had troubled the RDS. These included its relationship with the other Irish learned body, the Royal Irish Academy (at one point a merger had been proposed by the government, but the RIA did not cooperate), its relationship with the government, whose civil servants and administrators of the museum complex shared the society's Leinster House headquarters, and its relationship with George Francis Fitzgerald.

The irascible physicist, who was the honorary secretary of the society, had spent much time in 1886–8 assisting his uncle, George Johnstone Stoney, an educational administrator and spare-time physicist who was also a vice-president of the RDS, in rewriting the society's charter. The government largely accepted their draft, but a proposal by Fitzgerald that the society should create fellowships was less successful. Fitzgerald's intention was to encourage scientific work among the RDS membership and to emphasise its status as the 'Royal Society of Dublin'. The first fellows would have been those who were already fellows of the Royal Society. When the proposal failed to gain the approval of the society's membership at large in 1889, Fitzgerald resigned forthwith as secretary and had little more to do with the RDS.[2]

The society settled down to a quarter of a century of solid growth, concentrating on the agricultural activities with which it was

2   The episode is recounted with varying emphases in the histories of the RDS by Moss, in Berry (p. 310), White (pp. 130–1), and Charles Mollan in Meenan and Clarke (pp. 210–13).

primarily concerned. Its scientific mission was not neglected, however, and now that it had been freed from the burden of administering the institutions which had passed to the state, new avenues could be explored. It maintained a laboratory and chemical analysis service for its members run by Richard J. Moss, who seems to have mainly dealt with agricultural materials such as fertiliser and soil samples. Moss was also the society's registrar and therefore its chief administrator. It also continued to publish regular Scientific Proceedings and Scientific Transactions. These had been re-launched by Moss and Stoney in 1877 and indeed Stoney coined the word 'electron' in an RDS publication in 1891.[3] From 1890 the society began to make grants from its own funds in aid of scientific research which came to more than £600 over the next twenty-five years.[4] Its marine biology research programme, which started in 1890 and was taken over by the Department of Agriculture and Technical Instruction for Ireland in 1900, is described more fully in part II. It also founded the Royal Veterinary College of Ireland in 1895, and again this was transferred to the DATI in 1915.

The society not only encouraged science among academics – it ran (and still runs) a programme of public scientific lectures, and in 1897 triumphantly opened a new lecture theatre seating 700. Moss, writing in H.F. Berry's history of the society published in 1915, painstakingly details the air-conditioning, the organ, and the 'screen for lantern projections, which has an area of 340 square feet'.[5] Another initiative, probably inspired by John Joly, which promoted the society's scientific achievements was the award of the Boyle Medal, named in honour of Ireland's greatest chemist, to members of the society who had made notable contributions to science. The medal was first awarded in 1899, appropriately enough to George Johnstone Stoney, who had carried out most of his research in the RDS laboratories. For the medal's second award, in 1900, the RDS looked to the future rather

3   G. Johnstone Stoney, 'On the Cause of Double Lines and of Equidistant Satellites in the Spectra of Gases', *Scientific Transactions of the Royal Dublin Society* 4 (1891), 563–608.

4   These grants are listed by Moss, in Berry (pp. 370–1). The largest single items were £170 to C.J. Joly of Dunsink Observatory for a solar eclipse expedition and £100 towards the RIA's Clare Island Survey.

5   Moss, in Berry, p. 325–9.

than the past and honoured the 39-year-old Thomas Preston, professor of natural philosophy at UCD, who had written textbooks on heat and light and was carrying out fundamental research on the Zeeman effect (the magnetic splitting of spectral lines). Unfortunately Preston died suddenly less than a month after receiving the award.

John Joly, whom we have already met as a lobbyist for science within and outside Trinity College, was awarded the third Boyle Medal in 1911. He was also responsible for the RDS's last major scientific initiative, the Irish Radium Institute, in collaboration with the young Dr Walter Stevenson of Dr Steevens' Hospital. The use of radioactive material for the treatment of tumours had already been tried at various hospitals, but the standard practice of placing lumps of radioactive metal on the area to be treated was unsatisfactory for a number of reasons. Joly and Stevenson suggested in 1914 that instead the radioactive gas produced by radium, then known as Emanation but now called radon, could be pumped off a solution of radium bromide and supplied in glass capillaries to local hospitals. The RDS already had a supply of radium, and more was bought by public subscription, the two largest donors being Lord Iveagh and Sir John Purser Griffith. The equipment was designed by Moss, the registrar, and operated continually until it passed its equipment and remaining radium to a new state-aided Radiological Institute, set up by the Cancer Association of Ireland in 1947.[6]

Catholics in the RDS

The most recent history of the RDS draws attention to the slow but steady ascension of women members through the society's hierarchy (culminating a few years after the book's publication with the election of the first woman president in 1989). It is rather more coy on the issue of the progression of Catholics through the society's ranks.[7] Traditionally, the social make-up of the RDS had been very much biased towards the Ascendancy. This is demonstrated by its

---

6 This account is based on Joly's own 'History of the Irish Radium Institute, 1914–1930', in the Royal Dublin Society's *Bicentenary Souvenir* (1931), pp. 23–9. There are slightly differing accounts of the end of the Institute in White, pp. 202–3, and by Denis Crowley, in Meenan and Clarke, pp. 181–3.

7 Meenan and Clarke, p. 38.

succession of titled presidents after 1874 (when the lord lieutenant of Ireland ceased to hold the position automatically), who included the fourth Earl of Rosse (1887–92) and the Guinness magnate Lord Ardilaun (1897–1913). In 1929 the system was changed again to a single-term presidency of three years, the first of whom was John Joly. The society had started out with a membership which was overwhelmingly male, Protestant and rich, and its leading members inevitably reflected this.

This bias inevitably restricted the access of Catholics to the RDS's scientific activities. Under the 1889 statutes, the society had instituted a Committee for Science and its Industrial Applications consisting of those scientists on its council and a further nine members (increased to fifteen in 1915) elected by the society's membership at large. Of the fifty-three men who attended meetings of the committee between its inception in 1889 and its expansion in 1915, precisely four can be identified as having an Irish Catholic background.[8] Three were academic scientists at University College: Arthur W. Conway (who became the society's first Catholic president in 1941), Edmond J. McWeeney and Monsignor Gerald Molloy; the fourth, Sir Joseph McGrath, was an academic administrator.

From 1915, the reported election results for the additional members of the Committee for Science and its Industrial Applications include the rank order of the fifteen successful and the (usually) five unsuccessful candidates.[9] The electorate was the society at large. During the first eleven years of the new system, forty-eight candidates sought election. It can be seen in this summary of the election results that fewer than half of the Catholics who stood for election to the committee between 1915 and 1925 were successful, while more than three-quarters of the other candidates were elected at one time or another. The average rank of a

---

8   Those attending each meeting are recorded in the Minute Books of the Committee for Science and its Industrial Applications in the RDS archives. Information about the religious affiliations and backgrounds of the committee's members has been obtained from *Who's Who, Dictionary of National Biography* and similar sources.

9   There seem to have been only nineteen candidates in 1915, and eighteen in 1918, of whom four and three respectively were not elected. The full election results are given in Appendix 1.

Table 5.1: Summary of election results to the Royal Dublin Society's
Committee for Science and its Industrial Applications, 1915–1925

			(1)	(2)	(3)	(4)	(5)	(6)	(7)
Background	Irish	Catholic	16	7	9	71	41	52%	12.8
		Protestant	19	17	2	85	77	91%	9.0
	GB		11	7	4	56	42	75%	10.0
	Not known		2	2	0	6	5	83%	10.4
Institutional	Trinity	staff	13	10	3	45	33	73%	11.2
Affiliation	College	medics	4	4	0	22	22	100%	5.6
		total	17	14	3	67	55	82%	9.4
	State	Roy Coll Sci	8	7	1	50	45	90%	8.7
	Science	other Sci & A	6	3	3	28	23	82%	9.2
		other DATI	5	2	4	24	14	58%	12.2
		State Chemist	2	0	2	2	0	0%	18.5
		total	22	12	10	104	82	80%	9.7
	UCD		7	5	2	34	19	56%	13.5
	Industry		2	2	0	13	9	69%	12.9
All candidates			48	33	15	217	165	76%	10.4

(1) Number of candidates
(2) Number of candidates who were elected at some time between 1915 and 1925
(3) Number of candidates who were never elected between 1915 and 1925
(4) Number of candidatures (i.e. counting each time each candidate ran separately)
(5) Number of successful candidatures
(6) Percentage of successful candidatures
(7) Average rank of candidates (from 1st to 20th)
Source: RDS Election Results (see Appendix 1)

Catholic candidate in each election was slightly higher than thir-
teenth, more than two places below the average. Candidates from
UCD, or who happened to be UCD-trained state chemists, seem to
have been at a particular disadvantage. Even at this late stage, the
RDS was playing an important gatekeeping role by impeding
Catholic scientists from equal ease of access to the social status of a
seat on the society's main science committee.

It should be noted that although state-employed scientists
outnumbered those attached to Trinity College in the block of
fifteen elected members of the Committee for Science and its
Industrial Applications, the Trinity scientists found it much easier
to become members of the RDS council or vice-presidents of the
society, who were members of the committee *ex officio*. These also
included engineers such as Sir Howard Grubb, the telescope

manufacturer, Sir John Purser Griffith, and Samuel Geoghegan, who was responsible for the design of the RDS lecture theatre. In fairness, it should also be noted that at least one of the four or five scientific members of the RDS council elected each year after 1912 was attached to UCD, although none of the vice-presidents was until after 1925.[10]

The relative inability of Catholic scientists to gain recognition from the RDS must stem from the nature of the society's membership. The social tension between Catholics, who made up the majority of Ireland's population, and Protestants, who had traditionally made up most of the RDS membership, made it easier for a junior lecturer in Trinity or the state-owned Royal College of Science to get elected to the committee than for a professor in the National University. Three of the four most successful Catholic candidates taught at the Royal College of Science rather than UCD,[11] and were therefore less identified with the nationalist educational agenda which many RDS members must have viewed with suspicion.

Exile to Ballsbridge

The society's general failure to adapt to the changing social structure of Ireland almost led to its complete collapse around 1921. The crisis of the 1920s began several years before with George Noble, Count Plunkett, who was a member of the RDS and had been the director of the National Museum since 1907. His son, Joseph Mary Plunkett, was responsible for the military planning of the Easter Rising in 1916, and was subsequently executed by the British forces. (They were distant relatives of the Unionist politician Sir Horace Plunkett.) The count was sacked from his post in the museum, and rapidly became involved in the burgeoning Sinn Féin political movement, becoming its first elected MP. Inevitably, the Royal Dublin Society voted to expel him by 236 votes to 56 in January 1917.

---

10   A.W. Conway and H.J. McWeeney had already joined the RDS council by 1915. They were joined by H.J. Seymour from 1920, Felix Hackett from 1921 and Hugh Ryan from 1922. Joseph McGrath served as a Secretary from 1909 and as Vice-President from 1918 to 1921.

11   Pierce F. Purcell, Augustine Henry and Felix Hackett, respectively professor of civil engineering, professor of forestry and lecturer in physics and electrical technology.

By 1921, the political pendulum was swinging the other way, and Dublin Corporation, now much more sympathetic to Sinn Féin, banned RDS members from using the Dublin markets while Count Plunkett's expulsion stood. The society backed down speedily in April 1921 by remodelling the membership by-laws, which enabled it to re-admit the count and to broaden its membership base greatly as well. In the immediate aftermath of the Plunkett affair, the agricultural side of the society's council succeeded in placing the entire organisation under the control of a single manager, who was chosen for his skill at organising the agricultural shows which were the society's chief revenue-producing events. This meant retirement for Richard Moss, who had been the society's registrar and analyst for four decades, and was a severe blow to the society's scientific activities.[12]

Worse was to come. If the RDS had not gone to the trouble of designing a state-of-the-art lecture theatre in its central Dublin premises, it might be occupying them to this day. But the Anglo-Irish Treaty, which recognised the Irish Free State, signed in December 1921, also established Dáil Éireann as the parliament of the new state. In June 1922, Michael Collins, the head of the provisional government, requested that the RDS lecture theatre be made available for the sittings of the Dáil, and the Society duly obliged. As time passed it became clear that the government were not going to find suitable alternative premises for the Dáil and Senate, and the RDS vacated Leinster House entirely (with suitable compensation) and built a new headquarters at the site of its agricultural showgrounds at Ballsbridge.[13]

Moss's retirement and the move to Ballsbridge decisively shifted the society's activities from science to agriculture – although the Boyle Medal is still awarded, its scientific journals continued until 1985 and it is still the base for scientific lectures and educational projects such as the Young Scientist of the Year competition and the

---

12 White, *RDS*, ch. 15, pp. 173–88.
13 White, *RDS*, ch. 16, pp. 189–93. Senator Maurice Manning has suggested to me that the crucial lobby in favour of the Dáil and Senate decision to remain in Leinster House was a group of senators of Ascendancy background, motivated by the building's proximity to the gentlemen's clubs in Kildare Street and St Stephen's Green.

Youth Science events held in the summer. It also soon began the policy of celebrating the historical achievements of Irish science which has recently culminated in the publication of a series of monographs. The social barriers to Catholic scientists getting involved in the RDS dissolved as the society broadened its membership – particularly after 1925 – and as its investment of time and money in its scientific activities decreased. In 1941 Arthur W. Conway became the first Catholic to serve as president of the Society. Felix Hackett, a Catholic from Omagh, Co. Tyrone, the professor of physics at the Royal College of Science and then at UCD, succeeded in making himself indispensable to the RDS and served as its secretary from 1933 to 1953 when he in his turn became president.

Of the three Ascendancy institutions surveyed here, the scientific activities of the RDS had the least successful transition to the new regime after independence. This was only partly due to the circumstances of its enforced move of premises. Joly and Fitzgerald's plan to reinvent the society as a centre of Irish scientific effort contained the seeds of its own downfall. The society's Committee for Science and its Industrial Applications became not a microcosm of Irish science but a reflection of what the RDS members believed Irish science to be – a male, predominantly Protestant activity, dominated by Trinity and the Royal College of Science with a leavening of wealthy engineers like Sir Howard Grubb and Samuel Geoghegan. Bolting the science committee onto its agricultural activities only ensured that the society continually had to justify expenditure on its scientific activities to its membership, whose interests were primarily agricultural. And when the society's membership was broadened to include more Catholics after the move to Ballsbridge, the new recruits found a science committee which included very few of their number and which had resisted the very increase of emphasis on the society's agricultural activities which had attracted most of them to join it.

6

Royal Irish Academy:
successful transition

The Royal Irish Academy was founded in 1785 to promote the study of 'Science, Polite Literature and Antiquities', an outgrowth from previous eighteenth-century societies such as the Physico-Historical Society of Ireland, the Medico-Philosophical Society, and the Trinity-based Palaeosophers and Neophilosophers of the early 1780s. It had enjoyed a peak of activity in the middle of the nineteenth century as the focus both for the mathematical and astronomical research described in chapter 3 and for the study of Irish antiquities, in which the main figures were George Petrie, Samuel Ferguson and Sir William Wilde. Relieved of a major administrative burden, the antiquities collection, by the new National Museum in 1890, and also having repelled the RDS's attempts to encroach on its territory, the academy settled down to its role as an analogous institution to the Royal Society and the British Academy.[1]

---

1   *The Royal Irish Academy: A Bicentennial History, 1785–1985* (1985), abbreviated hereafter to *RIABH*, is the main source for what follows. The relevant chapters are 'The Main Narrative' (R.B. McDowell), 'Mathematics and Theoretical Physics' (T.D. Spearman), 'Experimental Physics' (T.E. Nevin), 'Astronomy, Geology, Meteorology' (G.L. Herries Davies), 'Chemistry' (E.M. Philbin) and 'Biology' (C. O hEocha). Sean Lysaght's Ph.D. thesis, 'Robert Lloyd Praeger and the Culture of Science in Ireland: 1865–1953' (1994), in particular ch. 7, describes Praeger's extensive activity within the RIA from his election in 1892 to his death sixty years later.

Uniquely among the institutions founded by the Ascendancy, the RIA was faced with and defeated a serious challenge to its status from Irish nationalists at the time of transition to independence. Its survival seems to have been the result of institutional reforms undertaken between 1890 and 1903, whose effects were to transform the academy from an Ascendancy institution to one whose composition more accurately reflected the different strands within Irish learning. Although this delicate consensus was strained by the events of 1916–22, the academy was able to adapt quickly to a changing political situation and to exploit divisions among its opponents. Learned societies in similar situations elsewhere did not surmount comparable challenges.

The Academy and science

The academy had been involved in the dissemination of learned knowledge since its foundation. Its *Transactions* were published from 1787 to 1917, and its *Proceedings* started publication in 1836 and continue to this day. In 1896 the *Transactions*, and in 1902 the *Proceedings*, were split into three sections: (a) mathematics, astronomy and physics; (b) biology, geology and chemistry; and (c) archaeology, linguistics and literature. From 1877 funds from a large bequest were used to produce a series of substantial Cunningham Memoirs, of which a majority seem to have been scientific rather than literary. Also in the 1870s, the academy began to provide grants for biological studies, and in 1885 and 1888 it sponsored the first of the marine biology expeditions which were to develop into the Department of Agriculture and Technical Instruction's Fisheries Branch.

The academy's activities in biological research were co-ordinated from 1893 by its Flora and Fauna Committee, which had a budget during the 1890s of 'from £70 to £130 annually out of an overall RIA budget of over £2000; concurrently the academy was spending between £300 and £400 annually on the production of its publications'.[2] The research sponsored by the committee at first consisted mainly of lists of the species to be found in various locations in Ireland, but greater projects were to come. A majority of the Flora

---

2   Lysaght, ch. 5.

and Fauna Committee's original members were employees of the Science and Art Department and this ensured that the profile of scientists in the employ of the state was enhanced within the academy. With the additional aid of the reforms of 1902, the number of state-employed scientists serving on the academy's Committee of Science rapidly increased. The scientific activities of the academy were no longer to be identified with Trinity College and Rowan Hamilton's quaternions.

The leading light among the influx of biological scientists was Robert Lloyd Praeger (1865–1953), born in Holywood, Co. Down, into a family with a tradition of liberal Protestantism and a strong interest in natural history.[3] From 1893 Praeger was an employee of the National Library, administered by the Department of Science and Art in South Kensington and later by the Department of Agriculture and Technical Instruction for Ireland. He also became a member of the Royal Irish Academy in that year. He served as the academy's librarian from 1903 and as president from 1931 to 1934. His scientific work was undertaken at weekends and during holidays. His first major work was the *Irish Topographical Botany*, published through the academy in 1901.

The Clare Island Survey, the Royal Irish Academy's most significant scientific initiative before the war, was led and directed by Praeger. It had been partly inspired by the success of a survey of Lambay Island off the coast of Dublin in 1905–6, carried out by the Dublin Naturalists Field Club under Praeger's direction, which discovered more than eighty species of flora and fauna new to Ireland and several which were new to science. In 1908, Praeger formed a committee to carry out a survey of Clare Island and the adjacent coast of Co. Mayo. The fieldwork for the survey was carried out over the summers of 1909, 1910 and 1911 and the results were published by the RIA between 1911 and 1915, sixty-eight papers which took up three whole volumes of the academy's *Proceedings*.

---

3   Apart from Lysaght's thesis and his shorter article on 'Science and the Cultural Revival: 1863–1916', in Bowler and Whyte, *Science and Society in Ireland* (1997), sources on Praeger include Timothy Collins, *Floreat Hibernia: A Bio-Bibliography of Robert Lloyd Praeger* (1985), and Praeger's own *The Way that I Went: An Irishman in Ireland* (1937) and *A Populous Solitude* (1941). Sally Montgomery, *Robert Lloyd Praeger* (1995), is a charming textbook for primary-school children.

Sean Lysaght has suggested that the motivation for these surveys was a desire on the part of Praeger and his natural history colleagues to find evidence for endemism in Ireland, endemism being the phenomenon whereby certain regions have plants and animals which are specific to them alone. Praeger himself had established that the number of species of plant present in Ireland is only about two-thirds the number present in Great Britain. This lack could be compensated for by arguing that the make-up of the Irish fauna and flora was unique to Ireland, with perhaps some elements surviving from before the last Ice Age as well as the main input from the neighbouring island since then. The links between natural history and cultural patriotism will be further explored in part II. Here we will note that the Clare Island Survey found little evidence for endemism on Clare Island. Indeed, its negative results undermined the usefulness of the biogeographical paradigm in biology and paved the way for a more ecological approach.

A reforming Ascendancy institution

The academy's pioneering work in the study of Irish antiquities and the recovery of the older Celtic culture laid the foundations of the Gaelic revival at the turn of the century. It is tempting to try and find an earlier tradition of support for Home Rule in the RIA in opposition to the more overtly Unionist RDS, as Norman McMillan claims to have done.[4] However, this is not really supported by the historical evidence. Throughout the nineteenth century scholars from the Ascendancy, in particular staff and graduates of Trinity College Dublin, dominated the academy. Many of its leading figures at the end of the nineteenth century had indeed been active in the Young Ireland movement in the 1840s. But most of these (for example Sir Samuel Ferguson and John Kells Ingram, who respectively served the academy as president from 1882 to 1886 and from 1892 to 1896) were less than enthusiastic about the new socially radical nationalism in their old age. The only overt Home Ruler to serve as president was the polymath Samuel Haughton, fellow of Trinity College, chemist, geologist, medic and clergyman, between 1886 and

---

4   McMillan, p. 113.

1891. The academy was no more balanced in religious terms than any other Irish institution of the nineteenth century. McDowell sums up the situation:

> [A] small minority of the members were Catholic, the overwhelming majority being Protestants, professional men or gentlemen with private means. Most of the Catholics who became members of the Academy had much the same social background as the other members and very similar views on many issues. Although only one of the eighteenth- and nineteenth-century presidents was a Catholic, from the [eighteen] forties one officer was usually a Catholic.[5]

There was certainly antagonism between the RIA and the RDS during the 1870s and 1880s, as both competed for an ever more closely scrutinised public funding and as the question of merging the two bodies was aired. These issues were exacerbated more by the fact that the scientific faction who controlled the RDS, led by George Johnstone Stoney, were also members of the academy but in a minority there,[6] than by any mutual antipathy between the institutions based on the wider issues of Irish politics. While the RDS was happy to be identified with unionism, the RIA preferred to present itself as a more pluralist body. Nonetheless, it remained Ascendancy in sentiment. In 1876 it rejected an application for membership from Isaac Butt, the founder of the Home Rule party, prompting a pained poetic rebuke from Samuel Ferguson.[7]

In the late 1890s John Henry Bernard, the academy's secretary, then a fellow of Trinity College and a future provost, promoted two important internal reforms that had far-reaching consequences. The first reform tightened up the criteria required for membership of the academy. Originally there had been no restriction placed on either the total number of members or on the number to be elected each year, with the result that both the quantity and the quality of new members were declining. Bernard successfully persuaded the council

---

5   McDowell, in *RIABH*, p. 50. The sole Catholic president was Sir Robert Kane from 1877 to 1882.

6   McDowell, in *RIABH*, p. 66.

7   S. Ferguson, 'To Mr Butt (On the result of the Ballot at the Royal Irish Academy, 13th November, 1876)', in *Poems* (1880), p. 167.

to recommend only twelve candidates per year for election after scru-
tinising their scholarly achievements, and convinced the academy
that no more than twelve should be elected (the number was
reduced to seven in 1917). Another new rule was introduced requir-
ing the two longest-serving members of each of the two committees
which constituted the council who were not office-bearers to retire
each year, so at least four of the twenty-one places on the council
were open at each election.[8]

The Council of the Academy consisted of its president, eleven
members elected to the Committee for Science and (from 1870) ten
elected to the Committee for Polite Literature. Although the outgo-
ing members of the council selected the slate of candidates for each
year's elections, they had to nominate twenty-two names for the
Committee for Science and twenty-one for the Committee for Polite
Literature. A good spread of candidates was thus ensured, and the
results of elections can be taken to reflect the strengths of different
lobbies within the academy. It has been possible to determine the
institutional affiliation of each of the eighty-three men who served on
the Committee for Science between the years 1890 and 1939.
Thirty-five were fellows or staff at Trinity College, and another two
were closely associated with the college; six were staff of the Royal
College of Science, eight otherwise employed by the Science and Art
Institutions, and two elsewhere within the department of Agriculture
and Technical Instruction; twenty were employed by UCD or other
colleges in the National University of Ireland after 1908 (including
two who were staff of the Royal College of Science before 1926);
three were professors at Queen's College/Queen's University,
Belfast; seven were amateurs with other sources of income; and a
Church of Ireland archbishop and the principal of the Veterinary
College complete the numbers.

Bernard probably did not anticipate that the effects of his two
reforms would be to dilute the Ascendancy bias of the academy.
The reform to membership made it more difficult for wealthy
Trinity graduates who wanted the social cachet of the letters
'MRIA' but were not well qualified academically to become

---

8   McDowell, in *RIABH,* discusses the reforms to membership, but not the
     changes to the council. For those, see the academy's *Minutes,* 1899–1902.

Figure 6.1: Membership of the RIA Science Committee, 1890–1934
(five-year averages by institutional affiliation)

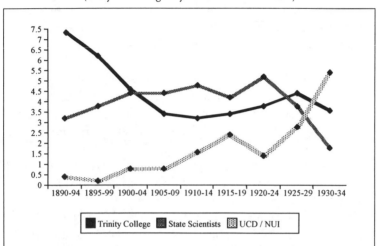

Trinity College = members of staff, plus Lord Rosse, the College's Chancellor, and C.R.C. Tichborne, a doctor who was also an examiner in the Trinity medical school.
State Scientists = staff of the Science and Art Institutions and two other employees of the Department of Agriculture and Technical Instruction.
UCD/NUI = UCD only to 1908, UCC and UCG from that date and former Royal College of Science staff from 1926. Others (QUB, the Queen's Colleges before 1908, and those who were independent of any institution) are not included here.
Source: RIA Science Committee election results (see Appendix 2)

members. But it also made it difficult for the council to refuse to support applications from suitably qualified scientific civil servants or lecturers at University College Dublin. The enforced rotation of seats on the council rejuvenated that body and also encouraged the newer members of the academy to participate in it. This is apparent from the recorded variation of the institutional affiliations of the members of the Science Committee for each five-year period between 1890 and 1934. The graph bears a striking resemblance to that produced by George Basalla to illustrate his theory of the phases in the diffusion of Western science.[9] The academy's Committee of Science was much less of a self-perpetuating oligarchy than was the equivalent committee of the RDS, and this pluralism gave it greater resilience when the crisis came.

---

9   Basalla, 'The Spread of Western Science', *Science* 156 (1967), 611–22, p. 612.

National Academy of Ireland vs Royal Irish Academy

Like the RDS, the Royal Irish Academy had a troubled time in the years before 1922. The central problem was almost identical – in the aftermath of the rising of Easter 1916, the academy had expelled John (Eoin) MacNeill, a professor of history at UCD.[10] Again, the political pendulum began to swing in MacNeill's favour, and in 1918 he was elected as an MP for the National University. In November 1919, a campaign to have him reinstated as a member of the academy was launched by Timothy Corcoran, SJ, the professor of education at UCD.[11] McDowell describes Corcoran as 'a large, solemn, stern man and a passionate nationalist who never shrank from controversy',[12] while E. Brian Titley has described his conservative vision of Catholic education at some length.[13]

Corcoran's first approach on behalf of MacNeill to the council of the academy, on 3 November 1919, was rebuffed on the slender procedural grounds that it was too late for new members to be nominated that year. Corcoran duly nominated MacNeill at the appropriate moment in February 1920, but the council did not recommend him for election in March, and he was not elected.[14] The discussion and votes within the council are not recorded, but they

---

10 MacNeill had nominally been in command of the Irish Volunteers, but the rising had been planned without his knowledge and he had tried to call it off at the last minute when he realised what was planned.

11 I have reconstructed the story of Corcoran's campaigns on behalf of MacNeill and of the National Academy of Ireland from his own papers, preserved in the UCD archives (LA20/3 and LA20/4 respectively), and from the relevant minutes of the Council of the Academy between 1919 and 1922, preserved in the RIA library, with some help from *RIABH*, pp. 81–5.

12 *RIABH*, p. 81.

13 E. Brian Titley, *Church, State and the Control of Schooling in Ireland 1900–1944* (1983), ch. 5 ('the New Order', pp. 90–100), critically assesses the contributions of both MacNeill and Corcoran to the educational ethos of the new Irish state after 1923.

14 The sponsors of MacNeill's nomination were four UCD professors (J.M. O'Sullivan, who was to succeed him as education minister in 1925; Osborn Bergin, a Gaelic scholar; Hugh Ryan, professor of chemistry; and Timothy Corcoran himself), a Protestant Home Ruler from QUB (R.M. Henry), the chairman of Convocation of the National University (M.F. Cox) and Count Plunkett, the former director of the National Museum, by now a minister in the revolutionary government.

cannot have been comfortable meetings. Several of MacNeill's colleagues at UCD were members of the Council, and Praeger had known him as a childhood holiday friend in Co. Antrim.[15] On the other hand MacNeill was now one of the leading figures within Sinn Féin and Dáil Éireann, which were closely linked to the terrorist campaign of 1919–21 that sought to overthrow the entire system of government to which the academy's Ascendancy members owed their allegiance and to which its civil service members (including Praeger) owed their jobs.

As a result, relations between the academy and nationalist Ireland worsened during 1920. A.W. Conway and Hugh Ryan, two professors at UCD who were involved in different ways in nationalist politics,[16] withdrew from the council before the March AGM. In the autumn, Corcoran got together a petition signed by forty-three members of the academy calling for MacNeill's reinstatement. About half of these were on the teaching staff of the NUI, only two were employees of the Science and Art Institutions (Timothy Hallissy of the Geological Survey and J.J. Buckley of the National Museum), and none were from Trinity College. Again, however, the council declined to nominate MacNeill for election to the academy, explaining rather weakly that it had not been quorate at the relevant meeting.

In March 1921, for the first time since 1905, no UCD scientist was elected to the council. In June 1921, Count Plunkett informed the council that he intended to propose that MacNeill's expulsion be rescinded at the next meeting. At this point the count, who had only recently been re-admitted to the RDS, was Minister for Fine Arts in de Valera's revolutionary Dáil government. King George V was about to make his famous appeal for peace talks which effectively ended the War of Independence, and Plunkett, MacNeill and de Valera appeared increasingly likely to become the real government before long. Although the council decided not to support the motion

---

15 Praeger, *Way that I Went*, p. 77.
16 Conway had actually contested the National University parliamentary seat as a Nationalist in 1918, losing to MacNeill. He was to contest it again, and lose to MacNeill again, in 1922. Ryan, running as a pro-de Valera candidate, also lost to MacNeill in the NUI constituency in 1923. See Brian M. Walker, *Parliamentary Election Results in Ireland, 1918–92* (1992).

by a narrow margin of seven votes to five, Plunkett had ensured that his supporters would turn out in sufficient numbers, and MacNeill was duly reinstated by the academy on 27 June. In a sign of change within the former Ascendancy, the motion's seconder was E.H. Alton, a future provost of Trinity who was to become a leading advocate of compromise among Southern Protestants.

MacNeill's supporters had achieved their first goal. But had the academy, with its royal patronage and curmudgeonly attitude to political progress, shown itself to be the kind of institution that any patriotic Irishman could work within? The academy was to exacerbate matters in November by rebuffing two overtures made to it by Count Plunkett in his ministerial capacity. Three days after MacNeill's reinstatement, a small gathering of interested parties was held in UCD which resolved:

> That it is desirable, in the interests of national Progress, to establish in Ireland a National Institute or Academy, for the Promotion and Publication of Original and critical Work, with Research and Invention, in the various departments of Science, Culture and Arts.[17]

A drafting committee was set up, consisting of ten NUI professors (including MacNeill and Corcoran) and a legal adviser. Science was represented in the persons of Hugh Ryan, Arthur Conway and T.P. Dillon (professor of chemistry at UCG). Two other members were scientists from outside Dublin (M. Conran from UCC and P. Browne of Maynooth) who did not in fact attend any of the meetings. By February 1922, the committee had decided to establish a National Academy of Ireland, with five divisions: (i) Irish Language and Literature, (ii) Mathematics and Physics, (iii) Letters, Philosophy and History, (iv) Biology and Anthropology, and (v) Fine Arts, and applied successfully for a grant of £100 from the Dáil.

By now the political situation was quite different to that of the previous June. The Anglo-Irish Treaty, signed on 6 December 1921, was ratified by sixty-four votes to fifty-seven in Dáil Éireann on 6

---

17 Corcoran papers, UCD LA20/4 (2). McDowell, in *RIABH,* wrongly dates this meeting 21 July; in fact it was held on 30 June 1921. The meeting consisted of five UCD professors, including three scientists (Ryan, J.J. Nolan and J.M. O'Connor), Corcoran and Osborn Bergin, the professor of early Irish.

January. MacNeill, who favoured it, was in the Speaker's Chair, but Plunkett was one of the large minority who supported de Valera. Mainstream Irish nationalism was divided, permanently as it turned out, and Michael Collins and Arthur Griffith, rather than Eamon de Valera, were now in control. The new climate provided the RIA with another chance for reconciliation. The day before the National Academy was to be launched, four of its supporters, including Ryan and Conway, received a letter from Felix Hackett, a physics lecturer at the Royal College of Science with a long career in committees ahead of him, on behalf of the Royal Irish Academy inviting them to become members of a committee which would:

> . . . prepare a statement as to the aims and achievements of the Academy as a National Institution with a view to placing its claims before the public and before the new Irish Government; it being also an instruction to the Committee to suggest any improvements as to the organisation of the Academy's work which they may think desirable.[18]

This committee had already been formed. It consisted of Hackett, E.C. Armstrong, the Keeper of Antiquities in the National Museum, E.H. Alton of Trinity College, Osborn Bergin, the professor of early Irish at UCD, and the RIA's office-holders. Corcoran must have been disappointed by Bergin's change of allegiance. Bergin had been one of the five who had issued the call for a 'National Institute or Academy' the previous June, although unlike Ryan and Conway he had also maintained his membership of the Council of the RIA over the previous year. Hackett too had originally expressed interest in the National Academy, but had now changed his mind.

Corcoran received a gleeful letter from Plunkett, gloating that 'it is something new to see the R.I.A. on its knees'.[19] On 2 March, after a meeting with Corcoran, the recipients of Hackett's letter decided that 'the scope of the investigations and reforms outlined in the letter from the RIA did not appear sufficient to the four signatories'. Corcoran drafted a formal reply for them declining to serve on the RIA committee, but the four chose not to use its two final para-

---

18 Hackett to Ryan, 25 February 1922, in Corcoran papers, UCD: LA20/8 (12).
19 Plunkett to Corcoran, 28 February 1922, in Corcoran papers, UCD:LA20/8

graphs. There would have been very little room for manoeuvre had the paragraphs below been included in a formal reply to the academy:

> We, and those with whom we are associated, have for a considerable period been convinced that the Royal Irish Academy has not fulfilled and does not afford (as it is constituted and directed) any prospect of fulfilling the purposes of 'a National Institution'. We do not conceive that it is possible by any limited process of 'improvements to the organisation of the Academy's work', to enable it to fulfil these aims.
>
> The radical changes which in our considered judgement would be necessary to attain the objects contemplated in the terms of ref[erence of] your committee would probably affect the entire structural organisation of the Royal Irish Academy; its standards for membership; the method of choosing members; the representative character of its Council and Committees, and its general relations to its Academic aims and to the Irish Nation.[20]

The next two weeks must have seen that room for manoeuvre being used, because at the AGM on St Patrick's Day all four recipients of Hackett's letter were elected to the council of the Royal Irish Academy. At the next meeting of the National Academy's drafting committee Corcoran recorded breathlessly that 'the relations of certain members of the Committee to the membership of Council RIA, and to membership of the RIA was [*sic*] informally discussed, as were also other issues of future policy'.[21] It appears that his campaign to undermine the RIA's credibility had been dealt a crippling blow. Had the RIA managed to get the four elected to its own council in order to put them in an awkward position? Or had Corcoran perhaps pushed his authority too far in the meeting of 2 March?

As the date for the launch of the National Academy drew closer, correspondence arrived for Corcoran accepting or declining membership of the new body. Most accepted, though there was least enthusiasm from those who had been invited to join sections I and III. Osborn Bergin, declining membership of the Irish studies

---

20 Corcoran papers, UCD LA20/8 (14), dated 3 March 1922. These two paragraphs are struck through.

21 Minutes of meeting of 20 March 1922, in Corcoran papers, UCD LA20/4 (3).

section, explained that he was committed to the RIA for the foreseeable future, and that:

> In any case the reasons given for founding an entirely new institution, instead of reforming & stimulating existing bodies, seem to me inadequate. Most of the sections will be international in character, and as for the first section, which I am invited to join, some of the names in your list strike me as irreconcilable with the 'exacting standards of membership' of the new Academy, and augur ill for the future efficiency of a body so constituted.[22]

On 10 April, the Royal Irish Academy published a full statement of its indispensable role in the promotion of Irish culture and the intellectual life of the nation. Naturally copies were sent to the government as part of its campaign to be recognised as the sole body competent for this role in the new state – and to ensure that its previous treasury grant should continue to be paid under the new regime. Meanwhile Dámh Ioldánach Éireann, the National Academy of Ireland, held its inaugural meeting on 18 May.[23] Forty-three of the eventual 106 members attended, and MacNeill, now the minister for education, was elected its president.[24] Two weeks later, MacNeill's government authorised payment of the usual treasury grant to the Royal Irish Academy but warned that it was made 'without prejudice to any change that may be decided upon at any future date'. On 22 June, the National Academy met again and elected boards for each of its five divisions.

Over the next few days the external political situation degenerated into crisis as the civil war between supporters and opponents of the Anglo-Irish Treaty broke out. For the next six months or so ordinary life could only be carried on under serious restrictions: the buildings around Leinster House, including the Royal College of Science, were closed for over a year. The National Academy had no chance of getting under way under such circumstances, particularly since the civil war bitterly divided its members but had a unifying

---

22 Bergin to Corcoran, 6 April 1922, in Corcoran papers, UCD LA20/8 (101).
23 McDowell has this meeting on 19 May (*RIABH*, p. 84).
24 McDowell counts thirty-two NUI staff, eight from Maynooth and twenty-nine other Catholic ecclesiastics, among the 126 members.

effect on the Royal Irish Academy. In early 1923, MacNeill – the National Academy's nominal president – presented two important papers to the RIA, and at that year's AGM J.J. Nolan – one of the National Academy's secretaries – was elected secretary of the RIA. A meeting of the National Academy was summoned in November 1923. How many of its members appeared is not known, but there were no more meetings.

The survival of the Royal Irish Academy is quite remarkable when one compares it to similar institutions elsewhere. Finland, like Ireland, gained independence in the aftermath of the First World War. It too promptly suffered a civil war, much worse than the Irish one; it too had – and still has – a fundamental cultural division, with a Swedish-speaking population roughly equal in proportion to that of the Protestant population of the newly independent Irish Free State. Among scientific societies in Finland, this period saw divisions along cultural lines appearing: the predominantly Swedish-speaking Societas pro Fauna et Flora Fennica, founded in 1821, was challenged by the more radical Finnish-speaking Vanamo society in 1896; the Finnish Society of Sciences, founded in 1838 and with a tendency to publish in Swedish, German and Latin, was opposed by the Finnish Academy of Sciences from 1908; and the Swedish-dominated Finnish Chemical Society, founded in 1891, found itself gradually losing ground to the Society of Finnish Chemists from 1919 with the two organisations formally severing links in 1927.[25]

While the failure of the National Academy of Ireland can be explained in large part by the outbreak of the civil war and the unpleasant personality of its progenitor, a comparison with the Finnish experience brings out some of the strengths of the Royal Irish Academy which would not otherwise be apparent. The most

---

25 My attention was first drawn to the similarities between Ireland and Finland by Jonas Jørstad, 'Nations once again', in *Revolution? Ireland 1917–1923*, ed. David Fitzpatrick (1990). For Societas pro Fauna et Flora Fennica/Vanamo and Finnish Society of Sciences/Finnish Academy of Sciences, see R. Collander (tr. by David Barrett), *The History of Botany in Finland, 1828–1918* (1965), pp. 107–11; for Finnish Chemical Society/Society of Finnish Chemists, see T. Enkvist, *The History of Chemistry in Finland, 1828–1918* (1972). For the more general background to Finnish history I have used Eino Jutikkala and Kauko Pirinen, *A History of Finland* (1979).

important difference between the Irish and Finnish situations was – and is – that in Ireland, religion is used as a signifier to separate two largely English-speaking cultural traditions, while in Finland, language is used as a signifier to separate two largely Lutheran traditions. Since the Irish language was never going to be used as a major vehicle for scholarly communication even by its enthusiasts, the RIA's use of English did not exclude Irish nationalists in the way that the older Finnish societies' use of Swedish excluded Finnish nationalists. While Protestant and Catholic scholars participated in the same discourse, Swedish speakers and Finnish speakers could not.[26]

Also, the RIA had age and experience on its side. Its eighteenth-century foundation preceded the revival of interest in Irish culture, for which indeed it had been largely responsible, and it possessed a number of unique cultural resources in its manuscript collection and its interest in the collections of the National Museum. In contrast, even the older Finnish learned bodies were founded at about the same time as the Finnish revival movement. Steve Yearley, who has also compared the two cases, concludes that Finland's political and administrative autonomy from Russia throughout the period of the nineteenth century was crucial, though he appears to be unaware of the splits within Finnish science over the language issue:

> Finland's earlier effective independence meant that the attempt to harness science to national aims was made at an earlier stage of its (science's) development. Equally the Grand Duchy came into existence before there existed an elaborate scientific community. The pressures which appear to have isolated the practice of 'high' science in nineteenth-century Ireland were therefore largely absent.[27]

---

26 That the language issue in Finland was a matter of personal political choice rather than family background is demonstrated by the story of the Neovius family, who produced a number of notable mathematicians throughout the nineteenth century. Edvard Rudolf Neovius was the professor of mathematics in Helsinki University from 1883 until he was made the minister for finance in 1900. The Finnish government of 1900–5 became deeply unpopular with nationalists, and when it collapsed Neovius had to emigrate to Denmark. His two brothers, Lars Theodor and Otto Wilhelm, both prominent mathematics teachers, changed their surname to Nevanlinna in 1906. See G. Elfving, *The History of Mathematics in Finland, 1828–1918* (1981), pp. 108–12.

27 Yearley, 'Colonial Science and Dependent Development: The Case of the Irish Experience', *Sociological Review* 37 (1989,) pp. 327–8.

Finally, it is interesting that the first of the Finnish splits occurred inside the natural history movement. Collander attributes this to a particular enthusiasm for their country's natural environment in the ideology of the Finnish Nationalists. In Ireland things went the other way. While the Catholic middle classes attended courses in Irish, the Protestant middle classes went on natural history outings, and the Royal Irish Academy had established itself as an institution which validated scholarly activity in both disciplines. This division was not impermeable – MacNeill wrote a historical chapter in the Clare Island Survey, for instance – but it was there.

7

The new regime

Ascendancy science was on the whole ill-prepared for the social revolution that occurred in Ireland at the start of the new century. The base of Irish astronomy was firmly established among the land-owning aristocracy, and as their ownership of the land disintegrated, so, quite literally, did most of the great telescopes. Trinity and the RDS both did their best to avert the coming storm, and inevitably science in both institutions suffered when the new order came to pass. The scientific eminence which Trinity had enjoyed in the nineteenth century eroded in the course of the twentieth, and the RDS took the hard-headed decision that it had to drop most of its scientific activities in order to survive.

The only Ascendancy-founded institution strong enough to survive a serious challenge to its status and to flourish in the new environment was the Royal Irish Academy. In 1925 its academic members were evenly divided among the three traditions which we have identified within Irish science: thirty-five were associated with Trinity College, thirty-three with the National University, and thirty-five with either the Royal College of Science or the other Science and Art Institutions.[1] The academy had successfully survived into the new era as a forum where the different strands of Irish learning could be balanced, thanks to the reforms initiated by J.H. Bernard a

---

1   McDowell, in T. O Raifeartaigh (ed.), *The Royal Irish Academy: A Bicentennial History* (1985), p. 78.

quarter of a century earlier which had weakened its identification with the Ascendancy.

Bernard himself survived until 1927 as provost of Trinity College. The penultimate chapter of McDowell and Webb's history of that institution has the title 'Low profile: 1919–1947'. They suggest that the college deliberately followed a policy of low visibility in the years after 1923, and that there were four possible courses for the ex-Unionist: to emigrate (though Plummer was the only one of the younger staff to do so); to accept the existence of the new regime but maintain allegiance to the old; to gradually and passively allow the new order to take its course; or to embrace it while seeking the best deal for Trinity and by extension for the Ascendancy tradition.[2] McDowell and Webb single out the physicist and later provost W.H. Thrift as a man of the last category, though the college had to wait until five years after his death to gain financial support from the Irish government.

The RDS reconciled itself to the new Ireland much more quickly than Trinity. Perhaps the best illustration of its newly found confidence is the programme for its bicentenary celebrations in 1931.[3] Eoin MacNeill's brother James, who had succeeded T.M. Healy as governor-general of the Irish Free State, heads the guest list, followed by the Irish cabinet. The official exhibition included sections devoted to agriculture, botany, chemistry, engineering, fisheries, geology, physics and zoology. On show were examples of Irish science over the previous two centuries and more, including Nicholas Callan's electromagnetic equipment from 100 years before and James Drumm's newly invented storage cell which was already being used for propelling suburban trains. However, this display also revealed that the RDS could no longer be considered a focus for scientific work.

The president of the RDS in its bicentennial year was the septuagenarian John Joly, who was still lecturing to unruly classes of engineering students in Trinity College. Like his college, he had as yet only reached an uneasy accommodation with the new status quo.

---

2   McDowell and Webb, *Trinity College Dublin, 1592–1952* (1982), p. 432.

3   Royal Dublin Society, *Bi-Centenary Celebrations, 1931: Official Handbook and Catalogue of Museum* (1931).

When the Cosgrave government, which had at least not fulfilled his darkest fears, lost power in 1932 he wrote to a correspondent:

> We are having an anxious time here – The Man in power – De Valera – is known to me. Personally I have nothing to say against him. He is a gentleman. But I fear he is obsessed with the 'Ireland-a-Nation' fad.[4]

Joly may have remembered de Valera from the latter's brief enrolment in Trinity College three decades before.[5] His own most significant investment in the scientific future had occurred a few years earlier when he persuaded Ernest Rutherford to take on a promising graduate student, Ernest T.S. Walton, for work in the Cavendish Laboratory in Cambridge. Walton and his colleague Cockcroft subsequently gained the Nobel Prize for their work in splitting the atom.[6]

Joly's term as president of the Royal Dublin Society coincided with Robert Lloyd Praeger's as President of the Royal Irish Academy. Praeger had taken advantage of a clause in the 1921 treaty to retire as National Librarian in 1923, and had spent the time since then writing a monograph on the genus *Sempervivum* and articles for the *Irish Statesman*. The natural history movement had faltered in the Free State. The *Irish Naturalist* had folded in 1924, to be replaced by the Belfast-based *Irish Naturalists' Journal*. Praeger himself however still had a role to play in the story of science in the new Ireland.

Could the Ascendancy scientific tradition have weathered the storm better? Roy Johnston has argued that a more sympathetic development of Irish nationalism might have been more open to Ascendancy scientists:

> One wonders how people like Fitzgerald and Tyndall would have evolved had the Irish separatist tradition developed along the lines laid down by [Thomas] Davis, rather than [Daniel] O'Connell; in this

---

4 Joly to Eva Rankin, 24 February 1933 (TCD MS 2312/316).
5 David Fitzpatrick, 'Eamon de Valera at Trinity College', *Hermathena* 133 (1982), 7–14.
6 Professor Walton recounted Joly's crucial role in his career in a conversation with me shortly before his death.

context, the political philosophy being secular and liberal-democratic, in the European tradition, the Queens Colleges would have become the National University, with Maynooth College as a constituent, Callan's work being taken on board. And if Parnellite Home Rule had broken through, the Protestant Parnell would have had no trouble in winning the support of the (initially Protestant) scientific elite for a policy of enrichment by recruitment from the Catholic intellectuals, via a non-sectarian 'laique' educational system.[7]

Johnston's argument is misleading in that it moves the spotlight off the failure of the Ascendancy scientists themselves to prepare for the coming transition of power, and onto the debate about the best way forward for Irish nationalism, which has always been his main concern. It is difficult to believe that Fitzgerald and Tyndall would have been susceptible to the nationalist political agenda no matter how liberal or democratic its pretensions, or that a viable secular nationalism was ever really on offer. The relations of nationalism and the Catholic church with science in Ireland will be the subject of another chapter.

We will conclude for now by observing that the fundamental problem for Ascendancy science was that it remained Ascendancy science, restricted to those at the upper end of the social scale and not particularly tied in to the interests of industry. By the end of the nineteenth century the state had taken over a number of scientific institutions founded by the Ascendancy, and the investment of public rather than private resources into scientific research was high on the political agenda. The next chapter will examine how this agenda was realised in Ireland.

---

7   Johnston, 'Godless Colleges and Non-Persons', *Causeway* 1, no. 1 (Autumn 1993), 38.

Part II

8

Administration scientists

The state-funded scientist in a colonial regime is in many ways the archetypal colonial scientist. After all, few colonies had developed a significant 'native' tradition of science by the end of the eighteenth century, as Ireland had. In most former colonies, the history of science is the story of the transfer of Western scientific skills from the coloniser to the colonised, and the first contact of the colonised with science is at the hands of the surveyor, whether astronomer, botanist, or geologist. In Ireland many of these surveying activities were initiated by Irish Ascendancy scientists, rather than at the behest of Britain, and then were annexed by the metropolitan government over the course of the nineteenth century.

Here we will examine the work of scientists in the direct employment of the Irish government between 1890 and 1930. Richard Jarrell asserts that the British state acquired an interest in controlling Irish scientific institutions in the mid-nineteenth century because science had begun to be perceived as a potential tool for national self-development at a time when Britain was hostile to Irish separatism. But by 1890, the situation had changed and the transfer of some powers of government from Westminster to Dublin seemed a very real and immediate prospect. The Department of Agriculture and Technical Instruction for Ireland (DATI), which controlled almost all of the state's scientific institutions, was set up in 1900 as part of a government policy of meeting some, but not all, of the Home Rulers' demands. Its first two vice-presidents, Horace Plunkett and T.W. Russell, were the first Irish politicians to have direct control of the Science and Art Institutions and of state-sponsored research in marine biology and agriculture.

The three most important of the Science and Art Institutions in Dublin were the Natural History Museum, the Geological Survey and the Royal College of Science. Institutional histories exist of both the Natural History Museum and the Geological Survey, but some new information is presented below. The material presented on the Royal College of Science is new. The other scientific activities funded by DATI will also be examined, including particularly the marine biology research that it sponsored.

The administration scientists, the second of the three major strands in Irish science which I have identified, were effectively those scientists employed in government scientific institutions rather than by universities or industry (and those employed by industry were very few). They were more likely to have moved to Ireland after an early career elsewhere than the Ascendancy or nationalist scientists, and they were also slightly more likely to be female. Three case studies of the use of Irish nationalist arguments during the 1900–20 period by administration scientists who were themselves English are presented. The dispute in each case was over the appropriate repository for one or more type specimens of uniquely Irish fauna. The three incidents perhaps illustrate MacLeod's concept of a 'moving metropolis' – the observation that the 'metropolis' to which a scientist working in a colonial environment may look is not necessarily always the colonising nation, but may be more local to the colony or even in a third country altogether.

After Irish independence in 1920–2, the state scientific institutions suffered from a loss of momentum. The staffing problems of the Geological Survey in particular will be illustrated. The Royal College of Science attempted to meet the new era by coming to a closer arrangement with Trinity College, but events outpaced these plans: the political controversy surrounding its closure during the 1922–3 civil war will be described. Finally, possible reasons for the slowing down of state scientific activity in Ireland in the years after independence are examined. The administration of state science in a newly independent Ireland is an important test case for the theory that Irish nationalism was somehow antagonistic to science.

State science in the nineteenth century

Richard Jarrell has already comprehensively covered the historical prelude.[1] He has explained how the Department of Science and Art in South Kensington took control of several scientific institutions in Ireland in the middle of the nineteenth century. Several, including the Botanic Gardens, the Dublin Museum of Science and Art, and the Royal College of Science, had originated in the activities of the Royal Dublin Society. The Irish wing of the Geological Survey had always been run from a London headquarters, but it had retained a certain amount of local autonomy. The Science and Art Institutions in Dublin also included the National Gallery, the National Library and the Royal Hibernian Academy (now the National College of Art and Design). These last will not concern us.

Jarrell's analysis is one of the few to have applied a colonial model to the history of any aspect of Irish science. He suggests that the island's British rulers were content to allow Irish science to develop independently until the early nineteenth century, but that, once the possibility developed that the Irish themselves might use science as a tool for national development, colonial policy dictated that the regime should try and control its use from the centre. He suggests that the centralised and well-established British Science and Art Department was pitted against the 'weaker, less integrated' Irish scientific institutions and absorbed them during the second half of the nineteenth century.

Jarrell's model of a centralising, strongly-organised system displacing a less centralised, less well organised system is a good one for the episodes he describes in the second half of the century. However, it does not account for the survival of Trinity College as a separate institution. Nor does it explain the eventual decentralisation of the Irish Science and Art Institutions to Irish control

---

1   Richard Jarrell, 'The Department of Science and Art and Control of Irish Science, 1853–1905', *Irish Historical Studies* 23 (1983), 330–47.

between 1900 and 1905, as part of the very agenda of national development which is supposed to be inimical to the interests of the coloniser. Jarrell admits that South Kensington's 'final withdrawal from Irish scientific life came about through a combination of bureaucratic reorganisation in London coupled with increased movement towards home rule'. The next few chapters deal with the consequences of that withdrawal.

9

DATI: the political background, 1895–1921

In 1900 most of the Science and Art Institutions in Ireland were detached from the South Kensington administration and placed under the new Department of Agriculture and Technical Instruction for Ireland.[1] DATI, as it was generally known, was effectively a ministry for agricultural development, set up as a result of successful lobbying by Horace Curzon Plunkett (1854–1932), a Unionist MP who became its first Vice-President (the department's nominal president was the Chief Secretary for Ireland). Plunkett's lobbying in favour of the creation of DATI succeeded in part because he was able to unite an impressive coalition of businessmen and Unionist and Nationalist politicians (the 'Recess Committee') behind his campaign. More importantly perhaps, it fitted in extremely well both with the Irish policy of the British government and with its attitude to the state's role in science.

Plunkett was from an Ascendancy background, a younger son of the Earl of Dunsany.[2] He entered Irish politics in 1889 after returning

---

1   The official history of the department is D. Hoctor, *The Department's Story* (1971).

2   Plunkett has attracted a number of sympathetic biographers, starting with his contemporaries E. MacLysaght, *Sir Horace Plunkett and his Place in the Irish Nation* (1916), and R.A. Anderson, *With Horace Plunkett in Ireland* (1935). More useful are Margaret Digby, *Horace Plunkett: An Anglo-American Irishman* (1949), and Trevor West, *Horace Plunkett, Co-Operation and Politics: An Irish Biography* (1986). See also Anne Buttimer, 'Twilight and Dawn for Geography in Ireland', in Bowler and Whyte, *Science and Society in Ireland* (1997).

from a ranching career in America, and threw himself with great zeal into the task of preaching, and as far as possible practising, the benefits of co-operative agriculture to the Irish farmer. Spurned by the Royal Dublin Society in his efforts, he founded the Irish Agricultural Organisation Society in 1894 to promote co-operation. After the general election of 1895 produced a decisive Unionist majority, he presided over an informal committee (called the Recess Committee because it met over the 1896 parliamentary recess) whose report recommended the institution of a Board of Agriculture as a separate department of the Irish administration. The Recess Committee's report suggested almost incidentally that the new department should include the Irish activities of the Science and Art Department, which were then being administered from South Kensington.[3]

This last point was really directed at the Department of Science and Art's control of technical education – which Plunkett and his colleague T.P. Gill (1858–1931) wanted to include in the new ministry – but there was little resistance to the transfer of the Science and Art institutions as well. There was however some scepticism outside the Recess Committee about the need for an Irish Department of Agriculture to have anything at all to do with technical education. The Catholic unionist Michael McCarthy, writing in 1901, called the Science and Art grant 'the best spent public money in Ireland' and expressed strong doubts about Plunkett's ability to administer it wisely.[4] George Moore may have been right to attribute this part of the scheme to empire-building on Gill's part: 'how Art [i.e. Technical Instruction] was gathered into the scheme I do not know, probably as a mere makeweight'.[5]

No case seems to have been ever made, not even by the hostile McCarthy, for a transfer of the Irish Science and Art grant, including

3   M.J. Clune, 'The Work and the Report of the Recess Committee, 1895–1896', *Studies* 71 (1982), 73–84. The transfer of the Science and Art Institutions is proposed in the *Report by the Recess Committee on the Establishment of a Department of Agriculture and Industries for Ireland* (2nd edn 1906) on pp. 104–5.

4   Michael McCarthy, *Five Years in Ireland: 1895–1900* (1901), p. 411. His encomium of the Science and Art grant is on pp. 137–8; ch. 28, pp. 402–26, is an attack on the constitution of DATI.

5   George Moore, *Hail and Farewell* (1976), p. 568. Ch. 9, pp. 564–77, of *Vale*, the third volume of *Hail and Farewell*, is a hilarious comparison of the activities of Plunkett and Gill in the operation of DATI with Flaubert's *Bouvard and Pécuchet*.

technical education, to a separate Irish education department. Yet the English and Scottish parts of the grant were devolved to the relevant boards of education in 1900, at the time of the creation of DATI, and in 1924 the Irish Free State did indeed transfer the entire Science and Art grant to the new department of education. So why was this arrangement not even suggested in 1900?

Apart from the positive force of Plunkett and Gill's initiative, an Irish demand for technical instruction to be included in the new department's operations which the government found it convenient to grant, there were political considerations which discouraged any British government from setting up an Irish education department *per se*. While the Catholic church would have bitterly opposed any scheme of state control of education, Protestants in both Ireland and Britain would have opposed any increase in local control of schools, which would inevitably have meant an increase in the power of the Catholic church. The National Schools scheme in the 1830s, an early attempt at compromise, had almost collapsed under attack from both sides, and Gladstone's government had been brought down in 1874 over the Catholic University issue. Technical education was however regarded by most as a non-religious matter and so separate provisions could be made for it with relative safety.[6]

Plunkett was in sympathy with the leader of the Conservative/Liberal Unionist coalition government, Lord Salisbury, and enjoyed cordial social relations with Salisbury's nephews Arthur and Gerald Balfour and in particular with Lady Elizabeth Balfour, Gerald's wife. The prime minister and his family were, for politicians, unusually interested in science issues. Salisbury himself had a small laboratory in his home, he and Arthur Balfour both served as presidents of the British Association, and the Balfours' sister Evelyn was married to the physicist Lord Rayleigh. Both Balfour brothers cut their political teeth in Ireland. Arthur served as chief secretary for Ireland from 1887 to 1891 and Gerald from 1895 to 1900. Both were associated with the Unionist strategy which became known as 'killing Home Rule with kindness'. This involved reforms such as the Congested Districts Board, set up in 1890 to improve economic

---

6   E. Brian Titley, *Church, State and the Control of Schooling in Ireland, 1900–1914* (1983), p. 9.

conditions in the West of Ireland; the Land Acts of 1891 and 1903, which empowered tenants to purchase land from their landlords; the local government reform of 1898, which instituted elected county councils; and the setting-up of the Department of Agriculture and Technical Instruction in 1899. The policy of 'constructive Unionism' was intended to complement vigorous suppression of political opposition – Arthur Balfour cracked down heavily on the nationalist 'Plan of Campaign', a co-ordinated rent strike on certain estates in 1887. It was also intended to pull the sting from nationalist grievances of unfair economic treatment by Britain and in particular to impress on the government's English supporters that there was an alternative to Home Rule. However, it provoked considerable dissatisfaction from the government's own supporters in Ireland, the Irish Unionists, who accused the Balfours and their allies of pandering to their traditional nationalist enemies.[7]

The creation of the Department of Agriculture drew particularly strong criticism from the Irish Unionists for a number of reasons. The Royal Dublin Society, which felt that it had a proud history of innovation in agriculture and, for that matter, in science, felt threatened by the creation of a new body to disburse large sums of government money for agricultural and scientific purposes. It is also clear from the tone of one sympathetic account that the society believed that Plunkett had marginalised it from the Recess Committee.[8] The pique suffered by the RDS's officers was augmented by the suspiciously democratic nature of the department, whose policies were to be scrutinised by a Council of Agriculture, most of whose members would be appointed by the new county councils.[9] Still worse was Plunkett's choice of T.P. Gill, not only a Catholic but also a former Nationalist MP, to be the secretary (administrative head) of the new department. The final straw was Plunkett's support of a bill to allow state funding of a Catholic University. F. Elrington Ball,

---

7   Andrew Gailey, *Ireland and the Death of Kindness: The Experience of Constructive Unionism 1890–1905* (1987), describes the Unionist policies during this period and the adverse reaction which they provoked from the government's own Irish supporters.

8   Terence de Vere White, *The Story of the Royal Dublin Society* (1955), pp. 168–72.

9   Andrew Gailey, 'Unionist Rhetoric and Irish Local Government Reform, 1895–99', *Irish Historical Studies* 24 (1984), 52–68.

the secretary of the Irish Unionist Alliance, stood against Plunkett in the 1900 general election with support from the Orange Order, the RDS and in particular from Lord Ardilaun (Arthur Edward Guinness). The resulting split in the Unionist vote lost the South Dublin seat to the Nationalists and ended Plunkett's parliamentary career.[10] Plunkett wrote reflectively in 1904:

> Now and again an individual tries to broaden the base of Irish Unionism and to bring himself into touch with the life of the people. But the nearer he gets to the people the farther he gets from the Irish Unionist leaders. The lot of such an individual is not a happy one: he is regarded as a mere intruder who does not know the rules of the game, and he is treated by the leading players on both sides like a dog in a tennis court.[11]

The RDS was perhaps the nearest Irish equivalent to the 'science lobby' of the time in Britain. English scientists seeking institutional innovation by the state were capable of exerting effective pressure on the Balfours, and also on R.B. Haldane and Lord Rosebery, the Balfours' pro-science counterparts in the Liberal party. According to Peter Alter, this pressure resulted in the creation of the National Physical Laboratory and Imperial College London.[12] By contrast, DATI itself was almost entirely Sir Horace Plunkett's idea. Its scientific innovations were limited to the setting-up of a seed-testing station as a joint venture with Guinness, the brewing company; the institution of an agriculture course in the Royal College of Science; and the setting-up of a permanent fisheries survey. Ireland's scientists had demanded none of these. The RDS was systematically excluded from the discussion, the Royal Irish Academy appears not to have been consulted, and indeed the reaction of the Royal College's professors to the news that it would be teaching agriculture in future seems to have been distinctly hostile.[13] The immediate

10 Alvin Jackson, 'The Failure of Unionism in Dublin, 1900', *Irish Historical Studies* 26 (1989), 377–95.

11 Horace Plunkett, *Ireland in the New Century* (1908), p. 64.

12 Peter Alter, *The Reluctant Patron: Science and the State in Britain, 1850–1920* (1987), ch. 3, pp. 138–90.

13 Reference in Plunkett's diary, quoted by Digby, *Horace Plunkett*, (1949), p. 20.

source of DATI's scientific programme was the Recess Committee's report,[14] backed up by Gill's research into the work of departments of agriculture in other countries.

T.P. Gill, whose appointment produced such controversy, had been a colleague of Plunkett's in the co-operation movement since it was founded.[15] His career in Nationalist politics had ended with the Parnell split, in which he had tried unsuccessfully to mediate. He had then become involved with the co-operative movement, and Plunkett made him editor of the Dublin *Daily Express* – much to the dismay of its traditionally Unionist readership.[16] His appointment as the department's administrative head had partly been intended to symbolise its cross-sectarian appeal by providing a religious balance to Plunkett, its political head, since Plunkett and both his successors as vice-president were Protestants.[17] Gill's influence on the department was to be greater in the long term than Plunkett's. He remained in charge of it throughout its entire life as an all-Ireland body, eventually retiring in 1923.

Plunkett's tenure of the vice-presidency survived the loss of his parliamentary seat in 1900 because he had succeeded in making himself appear indispensable to its continued operation. He attempted to return to parliament in a 1902 by-election in Galway city, standing at the request of dissident local Nationalists, but was unsuccessful.[18] To his own surprise, he retained the vice-presidency after the Liberals took power in late 1905. The new government made it clear that the work of the department was seen as 'non-political' and that Plunkett was still seen as important to its continued

---

14 See *Recess Committee Report*, pp. 76, 86 and 98–9.

15 Gill appears as an extra in the biographies of Plunkett mentioned above; there is no scholarly study of his career. His papers, many of them badly charred, are in the National Library of Ireland manuscripts department (MSS 13,478–526).

16 This is quite apparent from the sources in Patrick Buckland, *Irish Unionism* (1973), pp. 155–62, and confirmed by Moore's witty account in *Hail and Farewell*, pp. 564–77.

17 Anderson gives a rather jaundiced account of Gill's appointment which has become the received version in *With Horace Plunkett in Ireland*, ch. 8, pp. 104–16. He suggests that he (Anderson) would have been better suited for the position than Gill but that Gill's Catholicism got him the job.

18 The victorious Nationalist candidate was Colonel Arthur Lynch, of whom more later.

functioning. However, by this time, he had fatally undermined his own position with nationalists by including a trenchant criticism of the Irish national character in his book, *Ireland in the New Century*. A special commission was set up to enquire into the running of the department. Although its 1907 report recommended only minor changes, nationalists successfully exploited this opportunity to demand Plunkett's dismissal.[19]

The new vice-president of the department was T.W. Russell (1841–1920), the MP for Tyrone North since 1886.[20] Russell's career had begun as a journalist and temperance orator. He had first been elected to parliament as a Liberal Unionist. Under Salisbury he had been appointed an under-secretary to the (English) Local Government Board, but in 1902 he broke with the mainstream of Unionism on the issue of land purchase. Independent 'Russellite' candidates won two Ulster by-elections on the issue, and Russell represented tenants' interests at the 1903 conference which drew up the final land settlement. Like Plunkett, Russell was rejected by his former Unionist colleagues and lost his seat in 1910. Unlike Plunkett, he managed to do a deal with the Liberals and won a by-election in South Tyrone with nationalist support. As the only government supporter who was Irish, who was not bound by the nationalist pledge to reject public office, and who had ideas of his own about agriculture, he was the obvious man to appoint to the vice-presidency.

Plunkett's biographers have tended to be sympathetic to their subject and to have less time for Russell. One of them comments on the large number of letters Russell received to support candidates

19 M.J. Clune, 'Horace Plunkett's Resignation from the Irish Department of Agriculture and Technical Instruction, 1906–1907', *Éire-Ireland* 17 (1982), 57–73.

20 The only substantial treatment of Russell during his work for DATI is P.A. McKeown's unpublished Ph.D. thesis, 'T.W. Russell: Temperance Agitator, Militant Unionist Missionary, Radical Reformer and Political Pragmatist' (1991), in particular section 4 (pp. 240–322). Other aspects of Russell's career have been described by Alvin Jackson, 'Irish Unionism and the Russellite Threat, 1894–1906', *Irish Historical Studies* 25 (1987), 376–404, and J. Loughlin, 'T.W. Russell, the Tenant-Farmer Interest, and Progressive Unionism in Ulster, 1886–1900', *Éire-Ireland* 25 (1990). Russell describes his own political views and has some penetrating insights into the political situation of his time in *Ireland and the Empire: A Review* (1901).

seeking official positions, as if this were in some way a reflection on their recipient.[21] Russell successfully managed the affairs of the department for more than half of its life as an all-Ireland body, and was a vigorous legislator. His effort to distance the department from Plunkett's Irish Agricultural Organisation Society was an inevitable consequence of the close linkage that Plunkett had built up between the two during his time in office. Perhaps there was also the inevitable conflict of interest between state bureaucracy and voluntary organisations which had previously been instruments of government policy, the 'clash of systems' which Jarrell proposes for an earlier period. The split between DATI and the IAOS became magnified into a reflection of the deeper political divide in Ireland, Russell now being supported by Nationalists and the IAOS mostly by Unionists. Russell's activities were not limited to this dispute: his legislative activities were no doubt helped by his presence in parliament in contrast with Plunkett's absence. He moved swiftly to contain an outbreak of foot-and-mouth disease in 1912, and, more relevant for present purposes, he augmented the facilities of the department's fisheries branch and encouraged research and teaching in forestry by boosting the career of Augustine Henry.

The Development Act of 1909 created a Development Commission for the whole United Kingdom which had the power to make substantial grants towards the cost of scientific research, particularly research in the field of agriculture. Olby argues that the establishment of the commission marked an advance in state funding for science at this period which has not been sufficiently examined.[22] Olby himself does not examine the commission's impact in Ireland, other than to note that its bias towards funding projects at Cambridge University and Rothamsted inevitably reinforced existing centres of research rather than establishing new ones elsewhere. There are however a couple of peculiarly Irish wrinkles to the story of the Development Commission.

DATI saw itself as the agricultural research arm of the Irish government, and Russell in particular resented any intrusion onto

---

21 West, *Horace Plunkett*, p. 83. Hoctor, *The Department's Story*, pp. 50–92, gives a fairer comparison of Russell's tenure of the vice-presidency with Plunkett's.

22 Robert Olby, 'Social Imperialism and State Support for Agricultural Research in Edwardian Britain', *Annals of Science*, 48 (1991), 509–26.

what he perceived as his territory. His hand was strengthened by the requirement that all applications to the Development Commission must either originate from, or have first been referred to, the relevant government department, which was obviously DATI for all Irish applications. This enabled him to obtain substantial grants from the Development Fund towards improvements in fisheries and afforestation and for a veterinary research laboratory, and more especially for improving the seed-testing station at Ballinacurra, Co. Cork, and for setting up the chemical testing service that eventually became the State Laboratory.[23] At the same time, he went to considerable lengths to prevent a grant from being made to Plunkett's Irish Agricultural Organisation Society by the commission between 1911 and 1913, although he was eventually overruled by Augustine Birrell, the chief secretary.[24]

Russell's successor when he retired at the 1918 election was Hugh Thom Barrie (1860–1922), MP for North Londonderry and a leading Ulster Unionist.[25] Barrie had co-ordinated the Ulster Unionists' obstruction of the Irish Convention of 1917, one of the most earnest of the many attempts to get all Irish factions around the same table to thrash out a settlement.[26] His appointment may have been intended to help reassure Unionists that their cause was safe with the government, in the light of the election results which had left Sinn Féin with 73 of the 102 Irish seats. However, the end of British rule in most of Ireland was near. Barrie resigned in the middle of the Irish peace talks of late 1921 just after a Unionist conference in Liverpool had endorsed the negotiation process to which he was opposed, citing 'recent developments in the political situation'.[27]

Barrie's resignation does not seem to have perturbed anyone

23 Hoctor, *The Department's Story*, pp. 68, 80, 82, 159; DATI Reports after 1912.
24 McKeown, 'T.W. Russell', pp. 298–305, and DATI Reports 1911, 1912.
25 Lawrence W. MacBride, *The Greening of Dublin Castle: The Transformation of Bureaucratic and Judicial Personnel in Ireland, 1892–1922* (1991), p. 259, incorrectly states that Russell lost his seat; in fact he did not stand again.
26 R.B. McDowell, *The Irish Convention 1917–1918* (1970). See also MacBride, *Greening of Dublin Castle*, pp. 259–60, for critical press reaction to Barrie's appointment.
27 There can be little doubt that the Unionist conference was the turning-point for Barrie; cf. Lord Longford, *Peace by Ordeal* (1972), pp. 185–8.

unduly,[28] and his tenure of the vice-presidency has passed as unnoticed as his resignation by most historians of the period's administration.[29] He had been seriously ill for some time (he died five months later), and it is apparent from Gill's correspondence that he had not been very active as an administrator. The vice-presidency remained vacant until it was abolished as part of the Anglo-Irish Treaty, signed a few days later.

---

28 The only references to it that I have found are his farewell letter to Gill dated 21 November 1922, two days after his resignation, a copy of which is in the Geological Survey of Ireland archive, file 29, and in his obituary in the *Belfast Newsletter* (19 April 1922).

29 There is no mention of Barrie at all in either Eunan O'Halpin, *The Decline of the Union: British Government in Ireland, 1892–1920* (1987), or John McColgan, *British Policy and the Irish Administration, 1920–22* (1983); and MacBride, *Greening of Dublin Castle*, makes no mention of his resignation.

10

State-funded science in Ireland, 1890–1921

The Science and Art Institutions

Most of the Science and Art Institutions in 1900 formed (and still form) a block in the centre of Dublin, between Merrion Square, St Stephen's Green, and Trinity College. They were centred on Leinster House, the headquarters of the Royal Dublin Society from 1814 and also the administrative centre for the Science and Art buildings. Northeast and northwest of Leinster House respectively lie the National Gallery and the National Library. To the southwest is the main building of what was then the Dublin Museum of Science and Art (renamed the National Museum soon after DATI took it over). To the southeast were the two parts of the complex of most interest for present purposes: the Natural History wing of the museum (including also an annexe for its geological collections) and the Royal College of Science (which was completed only in 1910). A block or so away was the headquarters of the Geological Survey at 14 Hume Street.

Of the Science and Art Institutions, the Botanic Gardens at Glasnevin, north of Dublin, will be largely absent from my account.[1] They were established in 1795 by the (Royal) Dublin Society, and were originally managed by a keeper under the RDS professor of

---

1 This account is summarised from E.C. Nelson, *The Brightest Jewel: A History of the National Botanic Gardens, Glasnevin, Dublin* (1987), in particular chs. 9 and 11.

botany. Around the time of the establishment of the Royal College of Science under government control in 1867, this arrangement underwent some rapid modification. In 1878, the gardens passed to the control of the Department of Science and Art, and a clash of personalities rapidly developed between W.R. McNab (1844–90), the Royal College's professor of botany since 1872, and David Moore (1808–79), the curator (later director) of the gardens since 1848. Moore's successor after his death was his son Frederick (1857–1949), whose vision of the gardens was more as a horticultural institution than as a place for scientific research. McNab exploited the new relationship with the government to demand office and laboratory space in Glasnevin from Moore. In 1880 he succeeded in having himself made the gardens' scientific supervisor, conducting botanical research, mostly on his own initiative but occasionally on behalf of the government, and demanding the provision of plants for his classes in the Royal College.

The tension which continued between the younger Moore and McNab was resolved only by the latter's death in 1889. Thomas Johnson (1863–1956) succeeded McNab in the Royal College of Science, but was not granted his predecessor's post or privileges in Glasnevin. The focus of botanical research was shifted to the new buildings of the National Museum. McNab and his impact on Glasnevin were literally written out of the gardens' history – a historical article written by Sir Frederick Moore's successor as keeper in 1935 contains no reference to him at all.[2] The Moores however are commemorated in the title of the journal *Moorea*, published since 1982 by the gardens.

The Natural History Museum

The Natural History department of the museum had been opened in 1857 on the occasion of a Dublin meeting of the British Association for the Advancement of Science,[3] although the zoological part

---

2   J.W. Besant, 'Botanic Gardens: Origin, History and Development', *Journal of the Department of Agriculture* 33 (1935), 173–82.

3   The best account of the museum's history is C.E. O'Riordan, *The Natural History Museum, Dublin* (c. 1983); N.T. Monaghan, 'Geology in the National Museum of Ireland', *Geological Curator*, 5 (1992 for 1989), 275–82, gives a definitive guide to

of the collection alone filled this building. The geological collections were displayed on the ground floor of the main museum building for some years after it opened in 1890, but were then transferred to the annexe. There was also a botanical department on the top floor of the main museum building, the rest of which was devoted to the antiquities collection, mostly taken from the Royal Irish Academy, and the Irish Art and Industries collection from the Royal Dublin Society.

In charge of the Natural History department for most of our period was Robert F. Scharff (1858–1934), a zoologist who was on the museum staff from 1887 to 1921.[4] Scharff was born in Leeds of German parentage. He had studied in Edinburgh, London and Heidelberg and came to Ireland from the Marine Biological Station in Naples as assistant naturalist at the museum in 1887. He became curator later that same year. The title of the post was changed to keeper in 1890 when the main Museum of Science and Art buildings opened next door. Scharff became a zoologist of some note, specialising in what would now be called biogeography, the study of the geographical distributions of different species. His research led him to propose a number of land-bridges, now sunken, which had previously connected land areas now separated by seas or even oceans.[5]

When Scharff arrived in Dublin, the Science and Art Institutions were under the direction of Valentine Ball (1843–95), a member of the notable scientific Dublin family, who had worked in the Geological Survey of India before returning to Ireland.[6] Ball died in 1895 and was succeeded by Lt-Col G.T. Plunkett (1842–1922), a former

---

3　(*cont.*)　part of its geological collections. Stephen J. Gould briefly reviews the museum's history in 'Cabinet Museums Revisited', *Natural History* 103 (1994), 12–20. The museum's own records, at the time of writing, are in fairly good order; most correspondence seems to have been preserved, both of the Natural History branch (extremely well looked after in a series of filing cabinets) and of the museum as a whole (in need of conservation).

4　This summary of Scharff's career, and of his colleagues mentioned in the next paragraph, is based on the relevant entries in Robert Lloyd Praeger, *Some Irish Naturalists* (1949), with some necessary corrections based on O'Riordan, *The Natural History Museum,* pp. 28–32 and 60.

5　See Scharff, *European Animals* (1907).

6　Valentine Ball was the brother of Sir Robert Stawell Ball, the astronomer, and of Sir Charles Bent Ball (see ch. 4).

officer in the Royal Engineers. Plunkett retired in 1907, and was replaced as director of the museum (now separated from the other institutions) by the unrelated George Noble Plunkett (1851–1948), a papal count. The museum's name was changed to the National Museum in 1908. Count Plunkett's later political career has already been described. He was sacked in the aftermath of the Easter Rising, and Scharff filled in as acting director for the five years remaining until his own retirement in 1921.

Other scientific staff employed by the museum included G.H. Carpenter (1865–1939), a Londoner who came to the museum from South Kensington as an assistant in 1888 but moved next door to become professor of zoology at the Royal College of Science in 1904; A.R. Nichols (1859–1933), a Cambridge graduate who worked as a zoologist under Scharff and his predecessor from 1883, succeeded Scharff as keeper in 1921, and retired in turn in 1924; J.N. Halbert (1871–1948), a Dubliner who joined the museum as a technical assistant in 1892, made the transition to assistant naturalist after Carpenter's move in 1904, and stayed with the museum until retiring in 1923; Rowland Southern (1882–1935), who was born and trained as a chemist in Lancashire, took up zoology after moving to Dublin and was an assistant naturalist in the museum from 1906 but moved to the Fisheries Survey in 1911; Colin Selbie (1890–1916), a Scot who replaced Southern in 1911 and was killed at the Somme; and two women, Jane Stephens (1879–?), a Dubliner who became technical assistant and then assistant naturalist but had to resign on her marriage to Scharff in 1920, and Matilda Knowles (d. 1933), daughter of an Ulster antiquarian, who was employed on a temporary contract in the museum's herbarium from 1902 until 1933.

It is interesting to compare the experience of the Dublin museum with the case studies of natural history museums in Christchurch, Melbourne, Montreal, Buenos Aires and La Plata described by Susan Sheets-Pyenson.[7] She points to a general worldwide growth in interest in (and building of) museums up to 1890 (the year that the new Dublin buildings were opened). This was followed by a period

---

7   See Susan Sheets-Pyenson, *Cathedrals of Science: The Development of Colonial Natural History Museums during the Late Nineteenth Century* (1988), esp. ch. 2, 'Leaders and Followers: How Museums were staffed', pp. 27–44.

of precarious funding in the era of increasing local autonomy and an eventual shift to a programme of local studies rather than international collecting. The Dublin museum's funding and staff levels remained secure up to the change in the political situation in 1922. Scharff however was conscious of the political advantages to be gained for the museum, as Irish autonomy became more and more likely. As we shall see, he was prepared to challenge other scientists on those grounds.

Sheets-Pyenson also notes that support staff in colonial museums tended to be either graduates, imported from the metropolis, or local unqualified recruits, often from the lower social classes, who might well eventually achieve relatively senior rank despite their humble beginnings. In this context it is remarkable that only a minority of the Natural History Museum's staff (Halbert and the two women, Stephens and Knowles) were born in Ireland, and that they tended to start (and in the women's cases to remain) at a lower level than their better-educated English colleagues. Sufficiently qualified Irish recruits were scarce. Robert Lloyd Praeger, perhaps the most outstanding amateur scientist of his day in Ireland, remarked fifty years later that his failure to get the job which went to Carpenter after a competitive examination in 1888 was due to 'inadequate time for reading, and a disbelief in examinations – which I still hold'.[8] Praeger eventually used the security of a career with the National Library, on the far side of Leinster House, to support a passionate interest in natural history which enabled him to set the agenda for his professional colleagues.

The Geological Survey of Ireland

The Geological Survey was never based in the Leinster House site, and was the last of the scientific institutions to be taken over by DATI.[9] It had been instituted in 1845 after a number of private

---

8 Praeger, *A Populous Solitude* (1941), p. 188.

9 Largely because of the efforts of Gordon Herries Davies, the Geological Survey of Ireland is the best chronicled of the Science and Art Institutions. His *Sheets of Many Colours: The Mapping of Ireland's Rocks, 1750–1890* (1983), has already been superseded and brought up to 1984 by his latest work, *North from the Hook: 150 Years of the Geological Survey of Ireland* (1995). There is a good summary of the

initiatives to survey Ireland's rocks had collapsed as a result of funding problems or personality differences. Geologists had shared 51 St Stephen's Green with the Museum of Irish Industry and then the Royal College of Science until 1870, when new offices were opened in 14 Hume Street, where the survey was to remain for over a century. In 1887 the fieldwork for the primary survey was completed, and in 1890 the scientific staff was reduced from twelve to five.[10]

Because the survey was the Dublin branch of a London-based institution, and since it was supposed to be winding down in any case, it was not transferred to DATI in 1900. At this point the English and Scottish surveys had also finished their primary field-work, but were being allowed to undertake a survey of the effects of glacial drift. Lobbying by Irish press and politicians ensured that the Irish too were permitted to carry out a drift survey under G.W. Lamplugh. It became clear that the survey would continue as a lasting entity. In 1904, Lamplugh's imminent departure concentrated the minds of the Board of Education on the possibility of saving money by not appointing a replacement but instead transferring the six remaining scientific staff to DATI. This was done in 1905, and Grenville Cole (1859–1924), the geology professor in the Royal College of Science, was appointed the Irish survey's director.[11]

Three of the four geologists thus transferred were veterans of the

---

9   (*cont.*)   progress of geological work in Ireland up to the 1960s in the first chapter of John K. Charlesworth, *Historical Geology of Ireland* (1963). Edward Hull, *Reminiscences of a Strenuous Life* (1910), is the autobiography of the director of the Geological Survey of Ireland between 1869 and 1890. The survey's records have been conserved professionally, as the survey is a depository under the National Archive Act 1986. For the wider picture of geology in Ireland, see Patrick Wyse-Jackson, 'Fluctuations in Fortune: Three Hundred Years of Irish Geology', in Foster, *Nature in Ireland* (1998).

10   Herries Davies, *Sheets of Many Colours*, p. 228, and *North from the Hook*, ch. 5 ('Surgery 1890–1905', pp. 101–22).

11   This account is summarised from Jarrell, 'The Department of Science and Art', pp. 344–6. For Cole, see also Anne Buttimer, 'Twilight and Dawn for Geography in Ireland', in Bowler and Whyte, *Science and Society in Ireland* (1997), pp. 135–52.

previous reorganisation of 1890.[12] They were a close-knit group, all Irish recruits to the survey (unlike Cole who was English), and all but one was older than the 46-year-old new director. They were understandably perturbed at the appointment of the younger man and the disappearance of any remaining promotion prospects within the survey. The youngest, H.B. Seymour (1876–1954), resigned in 1909 to take up the newly created chair of geology at University College Dublin.[13] The others gradually retired, Alexander McHenry (1843–1919) in 1908, J.R. Kilroe (1848–1927) in 1913, and S.N.B. Wilkinson (1849–*c*.1927) in 1914. They were replaced by Timothy Hallissy (b. 1869), appointed in 1908, W.B. Wright (1876–1939), who had been working for the survey since 1901 but was only returned to Ireland in 1910, and two younger recruits, H.T. Kennedy (*c*.1890–1916) and R.L. Valentine (1890–1917), who replaced Kilroe and Wilkinson respectively. In 1908, before McHenry retired, the average age of the survey's geologists was fifty-four. After Valentine's appointment in 1914, it had dropped to thirty-three. Cole (and perhaps T.P. Gill as well) must have hoped that the new generation of geologists would revitalise the survey, but Valentine and Kennedy were both killed in the First World War, Valentine at the Somme in 1916 and Kennedy at Ypres the next year. Cole's own increasing arthritis prevented him from much activity in the last few years of his life, but the heaviest blow must have been Wright's transfer to England in 1921.

The Royal College of Science

The Royal College of Science was the last of the institutions to arrive at the Leinster House site.[14] It had originally occupied the buildings

---

12 There were two other professional staff, Richard Clark (1853–1933), the super-intendent of maps, who retired in 1918, and M.W. Gavin, the draughtsman, who retired in 1916. See also Herries Davies, *North from the Hook*, ch. 6 ('Recu-peration 1905–1924', pp. 123–51).

13 Geological Survey of Ireland archives, file 11.

14 There is a dearth of sources on the Royal College of Science. B.B. Kelham, 'The Royal College of Science for Ireland (1867–1926)', *Studies* 56 (1967), 297–309, concentrates on the college's history before 1890. Woodworth and Gorman, *The College of Science for Ireland: Its Origin and Development* (1923), is a necessarily sympathetic (and scarce) pamphlet written when the college was under the threat of imminent closure. Some of the college's records were preserved by the

of the Museum of Irish Industry in St Stephen's Green, and moved to palatial new buildings beside the Natural History Museum in the first decade of the century. These were formally opened by George V in 1910, on what turned out to be the last visit to Dublin by a reigning British monarch. The College shared its site with Plunkett's Department of Agriculture and the Local Government Board. A 1916 student magazine commented prophetically:

> The new wing of the College buildings which is approaching comple-tion will be occupied by the Department [of Agriculture and Technical Instruction]. The old houses in Merrion Street will then be taken down and a building erected to house the Local Government Board. The entire group will then consist of two government depart-ments and one College. This fact has stimulated a zoological contributor who forwards an account from an old Natural History of the Prairie Dog, the Owl and the Rattlesnake all living happily in one burrow. Recent investigations, however, have found that the Rattlesnake eats the Owl and compels the Prairie Dog to look for new digs. Our readers may allocate the titles.[15]

The Royal College of Science and the National Museum shared a common descent from the Royal Dublin Society and the Museum of Irish Industry, a government-funded institution set up in 1847 by the Irish chemist and scientific activist Sir Robert Kane (1809–90). The Royal College had been founded in 1867 by a government amalgamation of the RDS's scientific teaching staff with that of the Museum of Irish Industry, whose collection was handed over to the then Dublin Museum for Science and Art.[16] When it was estab-lished in 1867, the college had professors of physics, chemistry, applied chemistry, geology, applied mathematics, botany, zoology,

---

14 (*cont.*)   action of Dr Jim White when the buildings were evacuated by UCD in 1989, and are now in the UCD archives. A series of *The Torch*, a magazine produced by the Royal College's students, can be found in the National Library of Ireland. The report of the departmental committee which selected the site of the new buildings for the college on Merrion Street in 1897 was published as a command paper, C.9159.

15 *The Torch*, vol. 1, no. 1, January 1916, p. 2 (surviving copy in the National Library of Ireland).

16 Jarrell, 'The Department of Science and Art' (1983), pp. 339–42.

agriculture, engineering, and mining and metallurgy. In 1874, the agriculture chair was suppressed and the two chemistry chairs merged, and in 1899 the chair in mining was dropped. Plunkett revived the chair in agriculture in 1900 and increased the number of lecturers as well. A committee set up by Plunkett in that year to look at the future role of the college in DATI resolved that:

> [T]he object of the College should be to give advanced education in Applied Science of a character higher than that given in an ordinary technical or intermediate school without aiming at competition with existing University Colleges. Its main functions would be to train teachers for technical schools and for secondary and intermediate schools in which Science will be taught; to supply a complete course of instruction in science as applied to Agriculture and the Industrial Arts for general students; and by means of summer courses to give instruction in elementary Science on a scheme designed primarily for teachers in elementary schools.[17]

The 1901 committee also recommended that the Royal College staff and buildings should act as a resource for the department as a whole, and made specific requirements of several of its professors:

> The requirements of the Fisheries branch of the Department of Agriculture would be met, so far as the College and public funds are concerned, by the scientific assistance which it would be part of the duty of the Professor of Zoology to render to the branch at the request of the Department. The assignment to members of the Department's Fishery staff . . . of space in the Zoological part of the new buildings, would enable them to carry out their scientific work.
>
> We believe that it would be advantageous were all the geological collections in the museum to be in charge of the Professor of Geology, and further, think that in the event of the Irish work of the Geological Survey being transferred to the Department of Agriculture and Technical Instruction, it would be desirable that that work also should be placed under his charge.
>
> The Professor of Botany should continue to be in charge of the botanical collections in the museum and should act as the referee of

---

17 DATI Report, 1901–02, Appendix p. 137–8. The report is dated 17 January 1901.

the Department of Agriculture and Technical Instruction on botanical matters. Generally we consider that part of the duties of the teaching staff of the College should be to advise, when required, the Department on subjects which fall within their several branches of science.[18]

The recommendations of the committee were not remarkable. Thomas Johnson, the professor of botany at the Royal College, had been put in charge of the museum's botanical collections when they were set up in 1891. He seems to have devolved most of his museum duties to Matilda Knowles from 1907. Similarly, Grenville Cole, the Royal College's geology professor, was given control of the museum's geological collections in 1895. At first there was some friction because part of the collection was also supposed to be in the keeping of the Geological Survey. This eased when the survey was transferred to DATI in 1905 and Cole became its director. Indeed Cole's enthusiasm for his museum duties waned as the survey took up more of his time.[19] It is worth noting however that the Fisheries Branch of the department remained independent of the Royal College of Science, probably because G.H. Carpenter, the zoology professor, was more interested in land entomology than in marine invertebrates.

As can be seen from my graph (Figure 10.1), compiled from the Royal College's annual reports from 1890 until 1919, only a third of the college's students took the full A.R.C.Sc. I. (Associate of the Royal College of Science for Ireland) course before 1900. The destination of some of the other students becomes clear from frequent references, in the reports of the 1890s, to the success of former students in the Royal University's examinations. About a third to a half of these were women. The college claimed to have been the first of the Irish higher education institutions to admit women on an equal footing with men. Under the DATI regime before 1914, associateship

---

18 DATI Report, 1901–02, p. 139.

19 In early June 1910 there was a series of sharp exchanges between Count Plunkett and Scharff relating to the 'confusion' in the Natural History Museum's geological collections, which they clearly felt was caused by Cole's neglect of his duties at the museum (Natural History Museum Archives, 'Official Documents, 1906–1923' folder).

Figure 10.1: Students at the Royal College of Science, 1890–1919

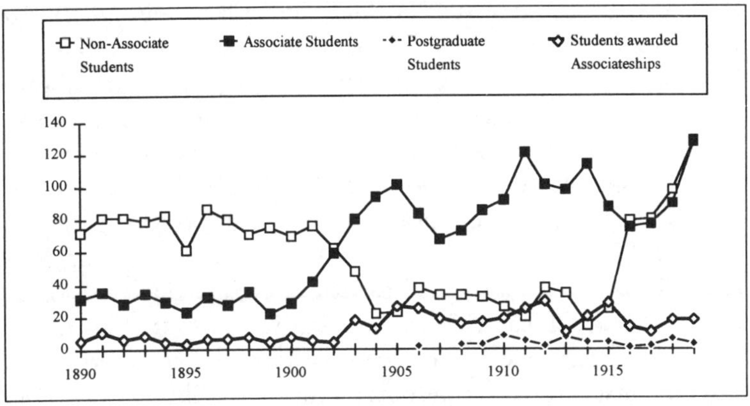

| -□- Non-Associate Students | -■- Associate Students | -◆- Postgraduate Students | -◇- Students awarded Associateships |

Source: Department of Science and Art and Department of Agriculture and Technical Instruction for Ireland annual reports.

students outnumbered non-associates by two to one on the whole. The proportion surviving through to the final examination remained about the same, and for the first time women qualified in some numbers as associates, most on the science teaching course – which seems for some to have been the first step of a scientific career. The standard three-year course was increased to four years in most subjects from 1910 and a new chair in forestry was added in 1912.

The number of students understandably fluctuated during the war, but increased rapidly thereafter. In 1918–19, the last year shown on the graph, there were 127 associate students and 131 non-associates. The next two years saw the college's largest ever intakes. In 1919–20, there were 214 associate students and 217 non-Associates, and in 1920–21 the figures were 236 and 183 respectively, as ex-servicemen whose education had been interrupted by the war returned.

The college could feel justly proud of its record as a research as well as a teaching institution: under Walter Hartley (1846–1913), its chemistry professor from 1880 to 1911, and his successor, G.T. Morgan (1872–1940), a flourishing research school was built up. Hartley was knighted at the opening of the college's new buildings in 1910 for his research on the absorption spectra of organic compounds; William Fletcher Barrett (1844–1925), its physics professor from 1874 to 1909, was also knighted;[20] and Belfast-born mathematics professor William McFadden Orr (1866–1934), like Hartley and Barrett, was a fellow of the Royal Society. Thomas Johnson's involvement with the museum's herbarium has already been noted. His reduced enthusiasm for it after 1902 may well have been due to his being put in charge of the new DATI seed-testing station. Carpenter and Cole published extensively on Irish animals and minerals; the professor of forestry, Augustine Henry (1857–1930), a Tyrone Catholic whose first degree was from Queen's College Galway, had carried out pioneering research on the flora of China before studying forestry at Nancy and Cambridge with a view to teaching it in Ireland.[21] Henry may have been the first Catholic to become a professor at the Royal College for some time. An unsigned paper (probably written by Plunkett) dated March 1906 states that the only Catholic then on the Royal College staff was Dowling, the registrar.[22]

This was quite apart from scientists who had left the college for greater things. The renowned populariser of astronomy Sir Robert Ball (1840–1913) had been its first applied mathematics professor.[23] Belfast graduate F.G. Donnan (1870–1956), a notable inorganic chemist, had lectured there in 1903–4 before moving to England.[24] And Alfred Cort Haddon (1855–1940), professor of zoology at the college from 1880 to 1900, progressed from an early specialisation in embryology to a keen interest in anthropology as a result of an expedition to the Torres Straits, and eventually left Dublin to

---

20 This may not have been entirely earned by his work in orthodox physics. Barrett's knighthood came three years after his retirement from Dublin in 1909; he was one of the founders of the Society for Psychical Research in 1882, and was its president in 1904–5. See H. Price, *Fifty Years of Psychical Research: A Critical Survey* (1939), pp. 48–51.

21 Sheila Pim, *The Wood and the Trees* (2nd edn, 1984), is a biography of Henry. A substantial amount of his correspondence survives in the archives of Royal Botanical Gardens at Kew, particularly in the bound volumes of correspondence to Sir William Thiselton-Dyer (v. 2, pp. 94–114) and Sir David Prain (v. 121, pp. 1763–99).

22 National Library of Ireland (T.P. Gill papers) MS 13484/2.

23 This was the brother of museum director Valentine Ball.

24 Donnan's papers in University College, London, carry very few reminders of his time in Dublin.

pursue his new subject very profitably in Cambridge.[25]

The Royal College was less successful at finding scientific jobs within Ireland for its own graduates, though some such as Mabel Wright became demonstrators in its laboratories or spent time in postgraduate research. An analysis of the 1923 list of recent graduates who had scientific publications to their name illustrates the dearth of employment opportunities for scientists who wanted to find work in Ireland.[26] Of the forty-nine individuals listed, only three were working in Irish industry, two in a Belfast textile company and one for a Cork brewery. It is possible that there may have been many others in the private sector who escaped the notice of Woodworth and Gorman, but such evidence as there is in the college's papers suggests that the majority of its graduates became science teachers. It should be noted that only half of the college's associateship students in the decade 1891–1900 were from Ireland.[27] Unfortunately, there is no evidence about the origins of later students. The Royal College does not seem to have fulfilled Plunkett's ideal of an engine which would revitalise Ireland's industrial potential. It did however manage to provide plenty of well-trained graduates for the Irish public sector[28] and to teach science to the next generation of schoolchildren. (Table 10.1)

Other state-funded activities

As well as the Science and Art Institutions, the department acquired at its foundation the Inspectors of Irish Fisheries, who had existed for decades in order to make and enforce fisheries by-laws.[29] Under

---

25 See A. Hingston Quiggin, *Haddon the Head-Hunter* (1942); also frequent references in H. Kucklick, *The Savage Within* (1991).

26 Woodworth and Gorman, *The College of Science for Ireland*, pp. 25–53.

27 See T.P. Gill's evidence in the *Second Report of the Royal Commission on University Education* (1902), Cd. 900, p.8, question 4034.

28 Hoctor, *The Department's Story*, p.133, gives an impressive list of names.

29 Most of the records of the fisheries branch seem to have been lost when the Department of the Marine, as it had then become, moved from Kildare Place in the mid-1980s. I am most grateful to Dr Christopher Moriarty of the department for sharing his unrivalled knowledge of its history with me. See also his 'Fish and Fisheries', in Foster and Chesney, *Nature in Ireland* (1997). Fortunately a series of letters from the Irish Fisheries staff to W.T. Calman, the assistant in the British Museum (Natural History) responsible for its Crustacea

Table 10.1: Analysis by employer of the graduates of the Royal College of
Science listed by Woodworth and Gorman, 1923

	Public Sector (not education)	Education	Private Sector	Total
Ireland	13 (12 in Free State, one in N.I. Ministry of Agriculture)	12 (5 in Royal College, 5 in NUI, 2 schoolteachers)	3	28
Britain	2 (both in Woolwich Arsenal)	9 (mostly in newer universities such as Leeds, Manchester, Sheffield, UCL, Birmingham)	8	19
Other	2 (Geological Survey of Egypt, Canadian Department of Agriculture)			2
Total	17	21	11	49

Source: Woodworth and Gorman, *The College of Science for Ireland*, pp. 25–53.

the Agriculture and Technical Instruction Act they were given
considerable extra funds for fisheries development and research.
Other scientific research, particularly in areas related to agriculture,
were also sponsored by the department, and it took advantage of
new sources of funds – both public (the 1909 Development Act and
later the DSIR) and private (Guinness) – as they became available.

Fisheries Research

Marine biology research off the Irish coast had begun with three
survey cruises carried out in the late 1880s by the Royal Irish
Academy. These were directed by William Spotswood Green
(1847–1919), then a Church of Ireland clergyman, but from 1890
one of the inspectors of Irish Fisheries and a member of the
Congested Districts Board. The photographs of the earlier RIA-
sponsored fishery research trips rather give the impression that
some prominent members of the academy, such as the three Ball
brothers, did not let their lack of expertise in matters marine deter

29 (*cont.*)   Collections, survives in the South Kensington archives. J. de C.
Ireland, *Ireland's Sea Fisheries: A History* (1981), almost completely ignores developments between 1900 and 1914.

them from participating in an expedition off the West coast of Ireland.[30]

For whatever reason, the government decided to fund future research through the RDS, which mounted further research trips in 1890 and 1891. The zoologist appointed to these was one E.W.L. Holt (1864–1922), an Etonian recently invalided out of the army after the Burma campaign (1886–7) who had dedicated himself with military efficiency to marine biology. He was appointed to run the Royal Dublin Society's newly established marine laboratory in 1898, and joined the Fisheries Branch when the laboratory was taken over by DATI.

The expansion of fisheries and its incorporation into the new department had also been one of the Recess Committee's demands in 1895. When the Inspectors of Irish Fisheries became the Fisheries Branch of DATI, Green was promoted to the new post of chief inspector, and Holt, as scientific adviser, was given a scientific staff of six, including S.W. Kemp and W.M. Tattersall, who both moved on to greater things in the scientific world, and G.P. Farran, whose career with the fisheries branch in its various subsequent incarnations lasted another forty years.

As well as its own marine biology research, the fisheries branch fitted into the programme agreed by the International Council for the Exploration of the Sea, despite the fact that, as Holt grumpily noted in the 1902–3 report, 'in the formulation of the scheme of international fishery investigation by the representatives of various continental powers and of Great Britain, it was apparently considered that the waters of Ireland need not be taken into account'.[31]

Whether this was due to neglect on the part of Whitehall or of Plunkett is not clear, but in the 1904 report Holt reveals that 'the Department has since indicated the formal adhesion of Ireland to the [international] scheme'.[32] He seems to be arguing for the rightful place of his own work on the international stage, rather than for equal representation of Ireland with Scotland – Scottish and English representatives both had votes on the council, but Holt

---

30 Praeger, *Some Irish Naturalists,* plates 16 and 17.
31 *Irish Fisheries Report, 1902–03, Part II* (1905), Cd. 2535, p. viii.
32 *Irish Fisheries Report, 1904, Part II* (1907), Cd. 3304, p. vii.

was happy to settle for observer status.[33]

More than half of the extra £10,000 which had been earmarked for fisheries development went on the upkeep of the fisheries protection vessel, the *Helga*. There were in fact two vessels of that name, the first being sold in 1907. Its replacement acquired lasting fame in Irish history when it was sailed up the Liffey at Easter 1916 and used to shell the surviving rebels out of the General Post Office. Even in peacetime much time had been spent pursuing foreign trawlers out of Irish waters. Besides the *Helga*s, the fisheries branch had access to the Congested District Board's *Granuaile*, thanks to Green's membership of the CDB.[34]

It is no exaggeration to describe the fisheries branch's initial research as of primary importance. A recent comprehensive bibliography of the Rockall Trough, that part of the Atlantic Ocean off the west coast of Ireland,[35] lists thirty-five papers published in the Irish fisheries reports between 1905 and 1913, a number hardly reached by the total of all papers published about the area in the following three decades. As well as surveying the species and abundances of fish off the Irish coast, Holt and his team managed to find many new species of marine crustacea.

Scientific publication itself was difficult to reconcile with the bureaucracy of government and the necessity for the department's publications to be presented as parliamentary papers. Holt complained that 'it is the devil's own job to get anything through the Secretary's office when one has, as in this case, to do it through the publication people'.[36] Despite his difficulties, five scientific reports were produced between 1901 and 1909 after a couple of years in the pipeline.[37]

---

33 Cf. also his contribution to *Minutes of Evidence Given before the Committee Appointed to Inquire into the Scientific and Statistical Investigations Now Being Carried On in Relation to the Fishing Industry of the United Kingdom* (1908), Cd. 4304, pp. 174–5.

34 *Irish Fisheries Reports, passim.*

35 J. Mauchline, D.J. Ellett, J.D. Gage, J.D.M. Gordon and E.J.W. Jones, 'A Bibliography of the Rockall Trough', *Proceedings of the Royal Society of Edinburgh, Section B – Biological Sciences* 88 (1986), 319–54. See also Whyte, 'Science and Nationality in Edwardian Ireland', in *Science and Society in Ireland* for more analysis of this and two other survey articles.

36 Holt to Calman, 10 August 1904 (South Kensington: Crustacea Correspondence).

37 The 1901, 1903, 1905 and 1907 scientific reports were all published two years after their date; the 1904 report was published in 1907.

Agricultural Research

As would be expected from an agricultural body, the department put considerable effort into testing different types of crop in order to improve the quality of Irish farming output generally. The annual reports detail experiments on such products as potatoes, barley, wheat, oats, tobacco, fertilisers, and livestock. The experimental plots seem to have been organised through the Irish Agricultural Organisation Society at first, and later by the agriculture committees of the county councils. The barley experiments were co-ordinated by the seed-testing station at Ballinacurra, Co. Cork, which was jointly owned by DATI and Guinness, the brewing firm.[38]

One of the spin-offs of the barley-testing scheme was its remarkable effect on the thoughts of William Sealy Gosset (1876–1937), an Oxford graduate who worked as a brewer with Guinness from 1899. Specifically, Gosset found himself faced with the problem of interpreting small samples in which the measures were not independent. This brought him into contact with the biometric school of Karl Pearson, and over the years from 1908, when his first paper was published in *Biometrika* under the pseudonym 'Student', he produced the definitive answer to the problem, now known as Student's *t*-test.[39]

Veterinary research was always of interest for the department, and in its first years of existence, Prof. Nocard, a disciple of Pasteur, was brought over from France to investigate a recurrent syndrome of calf mortality in the Limerick area. Using the best methods of Pasteurian research in the field, Nocard was able to identify a bacterium which carried the disease. Unfortunately he died before he was able to set up an Irish Pasteur Institute.[40]

Finally, the department was able to use the funds granted to it by the Development Commission to carry out extensive research into plant pathology. Seed-testing had been an early activity of

---

38 I am grateful to Professor Trevor West for information about Ballinacurra.

39 E.S. Pearson et al., *'Student', a Statistical Biography of William Sealy Gosset* (1990); see also S.R. Dennison and Oliver MacDonagh, *Guinness 1886–1939: From Incorporation to the Second World War* (Cork, 1998).

40 Besides the well-known work of Bruno Latour, *The Pasteurization of France* (1988), Anne-Marie Moulin, 'Patriarchal Science: The Network of the Overseas Pasteur Institutes', in *Science and Empires* (1992), gives an enlightening view of the agenda of Pasteur's disciples outside France.

the department under Thomas Johnson. This was expanded under G.H. Pethybridge (1871–1948) and P.A. Murphy, both of whom concentrated on diseases and infestations of the potato. G.H. Carpenter, the College of Science's zoology professor, carried out a good deal of entomological research for the department and published an annual list of injurious insects observed in Ireland.

The DSIR

The Committee of the Privy Council for Scientific and Industrial Research, later the Department of Scientific and Industrial Research, was formed in 1915 in order to redress a perceived lack of scientific research skills necessary for success in the modern world. Its first Advisory Council included J.A. McClelland, professor of applied mathematics in UCD, until his death in 1920 when he was replaced by Sydney Young, professor of chemistry at TCD. Another member of the council with Irish connections was Sir Charles A. Parsons, younger son of the third Earl of Rosse and brother of the fourth earl. The DSIR had substantial funds available for grants, but it is difficult to tell precisely how much of these were channelled towards Ireland. Certainly in the first report which lists papers published by recipients of DSIR grants towards postgraduate work, three of the twenty-five individuals mentioned had worked at the Royal College of Science in Dublin.[41] The DSIR also set up a committee of enquiry into the potential of peat as a source of power in Ireland, chaired by a civil engineer, Sir John Purser Griffith (1848–1938), and including the assistant secretary of DATI, George Fletcher, the chemistry professors of UCD and Trinity College, Hugh Ryan (1873–1931) and Sydney Young (1857–1937), and the professor of engineering at the Royal College of Science, Pierce F. Purcell (1881–1968), who wrote the eventual report.[42]

However, the DSIR arrived at precisely the wrong moment to make much of an impact in Ireland. Although other parts of the British Empire, notably Canada, were very keen to see DSIR

---

41 *DSIR Report, 1919–1920* (Cd. 905), App. IV, pp. 99–103.
42 Purcell, *The Peat Resources of Ireland* (1920).

transformed into a research-funding body for the Empire as a whole, the trend in Whitehall generally, and in Irish matters in particular, was toward greater autonomy and increased responsibility. In the event, funding for postgraduate scientific research in Ireland after independence came from only a very few sources: the National University awarded one travelling studentship every three years, and there was an annual award of one 1851 exhibition studentship for the entire Irish Free State.[43]

---

43 Information from Dr M.T. Bruck and the late Prof. E.T.S. Walton.

11

Administrators and nationalism, 1904–1920

How Irish were the scientists of the administration? To what extent did their location in Ireland determine the kind of science they were doing, and the way they thought about it? Most of them, particularly those at senior levels, were English, yet some individual English scientists working in Ireland were prepared to press the claims of Irish science in order to strengthen their own positions. In the three episodes which I examine below, the issue is the same: whether uniquely Irish specimens of marine crustacea, trilobites or foraminifera should be preserved in an appropriate Irish institutional setting or in a metropolitan, British base.

The political background against which these disputes took place changed considerably between 1903 and 1920. Irish Home Rule made its way onto the Statute Book in 1914, despite the opposition of the House of Lords and of the greater part of the population of Ulster. However, its implementation was to be delayed until the end of the war which had broken out in August 1914. Unionists and a majority of nationalists agreed to forget their political differences in the common cause of an allied victory which nobody expected to take four years. By 1915 the initial enthusiasm was beginning to wear off Ireland's war effort, and the discontent which culminated in the following year's rising was increasing. The British attitude seems to have been to reject the warning signs as mere whining. Had the Irish not been promised that they could have Home Rule, perhaps with some special provision for Ulster, once the Germans had been dealt with?

The years after the First World War ended brought the collapse of first traditional Irish nationalism and then British rule in most of Ireland. The political supremacy of Sinn Féin was consolidated in the 1920 local elections and it must have seemed certain to those scientists who worked for the state that Sinn Féin would form the future government of Ireland. It was at this point that G.A.J. Cole and R.F. Scharff, both English by birth but who had both been happy in earlier years to claim 'Irish' specimens for their Irish institutions, found themselves on opposing sides in a dispute over the proper repository for a dying geologist's unique collection.

The fisheries type specimens

The first of my three case histories of Irish nationalism in science involves R.F. Scharff of the National Museum and E.W.L. Holt of the fisheries branch of DATI. In 1904 and 1905, Holt was debating what should be done with the increasing number of specimens of marine crustacea which he and his colleagues were collecting from the seas off Ireland. Many were creatures which had not been previously described by scientists. A paper by G.P. Farran in the 1902–3 Fisheries Report lists thirteen new species (although Farran subsequently withdrew his claims to some), and a paper by Holt and W.M. Tattersall in the same report lists thirteen more.[1] In particular Holt was concerned about the fates of the type specimens (those specimens from which the official description of a species is drawn) of the new species being discovered. He shared his worries with W.T. Calman, the assistant in South Kensington in charge of the crustacea collections, in October 1904:[2]

---

1   The papers are appendices to the *Irish Fisheries Report, 1902–03, Part II* (1905), Cd. 2535: G.P. Farran, 'Report on the Copepoda of the Atlantic Slope off Counties Mayo and Galway', Appendix II, pp. 23–52, and E.W.L. Holt and W.M. Tattersall, 'Schizopodous Crustacea from the North-East Atlantic Slope', Appendix IV.i, pp. 99–152.

2   All the following letters are to be found in the Crustacea Correspondence in South Kensington.

I am most anxious not to have the question of disposal of material raised. If it is raised I may some day be compelled to hand over types of new species to our own museum, where they would be useless.[3]

Of course I want the types to remain in the British Museum, which is the proper place for them, but I don't want the fact advertised when they are solitary specimens.[4]

Holt's dilemma was clear. His instinct was that the British Museum should be the repository of type specimens, not just for Britain but for the rest of the Empire if not the world. However, he was aware that Scharff's stock was rising in Ireland, and was alert to the prospect that Scharff might demand that the type specimens of new Irish species be deposited in Ireland's National Museum. Indeed, a year later, this came to pass:

Scharff, Ph.D., is organising a campaign against me for presenting Irish types to the B.M. He daren't move in the matter personally, so he is putting in some outsider to call attention of Dept to this scandalous transaction and has kindly offered, if I will promise never to repeat the offence, to ward off the attack. I have told him that my hand has generally been strong enough to keep my head from worse dangers than this particular feud is likely to bring on me, and have suggested that his friends might be more profitably engaged in agitating to get his place made into a more fit receptacle for valuable specimens. The ways of the Irish patriot of Teutonic extraction are truly edifying.[5]

Holt adds in a postscript to this letter that he had just been made a council member of the Marine Biological Association, the main marine research body for England and Wales and by extension the British Isles. If he found a new species in British waters, doing research paid for by the British taxpayer, he was going to send the type specimens of the new species to the British Museum. Scharff, on the other hand, argued that they were Irish crustacea, discovered by a branch of the same Irish department which ran the National Museum.

---

3   Holt to Calman, 4 October 1904.
4   Holt to Calman, 12 October 1904.
5   Holt to Calman, 12 October 1905.

It seems that Scharff lost the argument on this occasion. The National Museum's catalogue of marine crustacea[6] claims that it possesses type specimens of only seven species, two of which were discovered in 1896 and the others after 1910 when Holt perhaps had relented. Various specimens of the species which were discovered by Holt and his colleagues between 1900 and 1910 are listed but none is described as a type specimen, so we must assume that the types were indeed sent to the British Museum.

The whole affair coincided with the 'Devolution Crisis' of 1904–5, when the Earl of Dunraven, with the active support of the most senior Irish civil servant, Sir Antony MacDonnell, attempted to persuade the Unionist government to adopt a scheme which fell far enough short of Home Rule to leave nationalists unsatisfied but was sufficiently innovative to infuriate Unionists. It is interesting to compare the worried tone of Holt's letters to Calman the month after the devolution crisis first broke in September 1904 with his more usual bombast the following year, some time after the chief secretary, George Wyndham, had brought the crisis to an end by resigning.[7]

The Tyrone trilobite

A very similar dispute arose between geologists in Ireland and Britain in 1915, this time over a number of fossils including the type specimen of a trilobite originally discovered in Co. Tyrone around 1840.[8] One of Grenville Cole's distant predecessors in charge of Irish geology was Colonel Joseph E. Portlock, an officer of the Ordnance Survey. Portlock suffered generally from an inability to see the wood for the trees. He completed a survey of half of Tyrone in the time he had been allowed for surveying the whole of Co. Derry, operating from the army barracks in North Queen Street, Belfast. Not surprisingly he was eventually transferred to other

6   C.E. O'Riordan, *A Catalogue of the Collection of Irish Marine Crustacea in the National Museum of Ireland* (*c.*1969).

7   O'Halpin, *Decline of the Union*, pp. 44–51.

8   The material for this case study is taken entirely from the British Geological Survey archive file GSM 2/409. Herries Davies refers to the incident briefly in *North from the Hook*, p. 123.

duties in Corfu. However, Portlock had managed to amass a substantial collection of fossils – unlike his immediate superiors he had realised the value of palaeontology for dating strata.[9] Some of his fossils were transferred to London from the Museum of Irish Industry collections at some point before 1870.

In June 1915, the Board of Education in England received a request from Gill, on behalf of DATI, that Portlock's fossils be returned. Attached was a memorandum from Cole, putting the case for returning the Irish fossils to their former home:

> In spite of its peculiar significance in the history of geological research in Ireland, [Portlock's collection] was at some date about 1870 transferred to the Museum of Practical Geology in London. The fossils are there 'amalgamated with other collections', and are distributed throughout the general collection of fossils of the British Isles. Numerous specimens described by Portlock appear in 'A catalogue of Cambrian and Silurian fossils in the Museum of Practical Geology' issued for H.M. Stationery Office in 1878.
>
> I understand that certain specimens have escaped from this collection into the teaching-collection of the Imperial College of Science . . . and that Portlock's type specimen of a trilobite, *Nuttainia hibernica*, from Tyrone, has been re-discovered with the original Ordnance Survey label attached to it. Now . . . it would seem highly desirable if the Department could approach the Board of Education in London, with a view to the return of the Portlock fossils to the collections of which they originally formed a part. It must be remembered that their removal may have seemed natural when the Geological Survey was administered from London as a whole; but this is the only case, so far as I am aware, when a distinctively Irish collection was so transferred.[10]

This caused some consternation in the British geological survey. Anxious memos flew for several weeks between its director, A. Strachan, and F.G. Ogilvie, permanent secretary to the Board of Education. They eventually concluded that it would be completely impractical to separate out the fossils from any one particular source in the Museum of Practical Geology's collection, particularly since

9   Herries Davies, *Sheets of Many Colours*, pp. 95–106.
10   Undated paper in GSM 2/409, dated receipt 15 June 1915.

the collection had probably been divided up long before 1864. A final note added that:

> the escape of the trilobite to which Professor Cole refers dates from a time when the Collections at Jermyn Street were used for teaching purposes in the Royal School of Mines then carried on in the same building. These Collections have not been used in this way for thirty-five years.[11]

This last barb was particularly intended to forestall any further efforts on Cole's part. Ogilvie, while going through the records, had discovered with glee that Cole himself had been employed as a demonstrator in Jermyn Street at the time of the trilobite's 'escape'.[12] 'The more one looks into it,' he commented to Strachan, 'the more does Prof. Grenville Cole's memorandum suggest the mode of presentation adopted in the Notes which the German Foreign Office sends to Washington!'[13] Cole's rebuff over the Portlock collection seems in keeping with British policy in general. In the midst of a world war, and at a time when traditional Irish nationalism was at a low ebb, there was little political mileage to be gained from trilobites.

The Wright foraminifera

My third case study concerns the foraminifera collection of Joseph Wright (1834–1923), which actually brought Cole and Scharff into conflict with each other. I have reconstructed this from the archives of the South Kensington Natural History Museum on the one hand and the National Museum[14] on the other. Joseph Wright was a

11 Typed memo from Ogilvie in GSM 2/409, dated 27 July 1916.
12 Handwritten memo from Ogilvie to Strachan in GSM 2/409, dated 16 July 1915.
13 Two decades later in 1936 the National Museum of Ireland and the Irish Geological Survey discussed the possibility of returning the Portlock fossils to their mother country. The Geological Survey informed the museum that Cole's attempt to do so had been unsuccessful, and that furthermore all the Portlock fossils had been collected in the six counties which remain under British rule. They remain in London to this day. See Dublin Natural History Museum, 'Incoming Correspondence'.
14 South Kensington Archives, Keeper's Correspondence, 1920; Dublin Natural History Museum, 'Official Correspondence' and 'Incoming Correspondence' drawers.

Quaker from Cork who made a successful career as a grocer in May Street in Belfast. Wright had become one of the world's leading experts on foraminifera, and was consulted by Canadian and Australian provincial governments. Robert Lloyd Praeger reminisced of him affectionately:

> A more kindly enthusiast than Joseph Wright never lived. I remember one occasion on which his self-restraint and benevolence were put to a severe test. In a dredging sent to him from – I forget where – he discovered a single specimen of remarkable novelty – the type of a new genus of Foraminifera. He mounted it temporarily on a slide – neglecting to put on a protective cover-glass, for he was a careless manipulator – and at a conversazione of the Belfast Naturalists' Field Club held immediately afterwards he showed it to J.H. Davies and others. Davies was a fellow-quaker, an ardent bryologist, a man of singular courtesy, a neat and skilful microscopical expert. Seeing that the slide was dusty, and not noticing the absence of the usual cover-glass, Davies leisurely produced a silk pocket-handkerchief and, before the horrified eyes of the owner, in a moment ground the specimen to powder! But Wright's self-restraint stood even that test. He gasped, and his face went white; but he uttered no word of reproach.[15]

By 1920, however, Wright was well into his eighties and in need of constant care. His relatives decided that the only thing to do was to sell Wright's lifetime collection of foraminifera, and approached the British Museum in February 1920 through William Swanston, another Belfast naturalist.[16] Grenville Cole had somehow heard that the approach was being made, and on 7 July 1920 wrote to Harmer, the British Museum's keeper of Natural History:

> I hear that you are considering the acquisition of the Wright collection of foraminifera for the British museum. I presume that some sample has been sent to you. Wright himself was a wonderfully careful manipulator, and some of his type-slides are veritable gems.[17]

---

15 Praeger, *Way that I Went* (1937), pp. 166–7.

16 Keeper's Correspondence 1920/72, Swanston to Harmer, 10 February 1920.

17 Rather different from Praeger's characterisation above! Perhaps Wright had learnt from experience to be more careful. Or perhaps Praeger had improved his anecdote with the passage of time – he was himself in his seventies when *Way that I Went* was published and the incident with Wright, Davies and the slide must have taken place at least four decades earlier.

It is very hard to value what are practically unique things, each specimen being selected by an expert, and I have been asked by old
friends in Belfast if I can give a fair idea. It strikes me that you may
be able to make an offer, since it is a bargain direct between scientific
men, and it is so difficult to asses the personal selection and critical
knowledge that was involved. 1725 slides at 1/6 makes a good sum; I
have no idea what the family would expect. The whole, if I judge
Wright's work aright (I love the old man, but must not be biassed
[*sic*] that way) is worth £150 to the nation. Is such a suggestion at all
improper at the present time? The leisure of a lifetime must be in the
collection, stored up for future workers. [inserted: Please do not send
my proposal to Belfast!] If you have any views, I can hand them on
or not, exactly as you wish. Luckily, I have no notion whatever of
the expectations of the Swanston-Wright combine.[18]

Cole added in a revealing postscript:

I am glad that the collection should find a home in the British Feder-
ation. Dublin is no secure place for Commonwealth treasures now.

The British Museum was feeling a post-war spending pinch, but
managed to contact an amateur English geologist, Edward Heron-
Allen, who knew and admired Wright, and was also an expert on
foraminifera. Heron-Allen was very keen to make an offer for the
whole collection:

Of course Joseph Wright's collections *must* come to the B.M. if not
direct, then through me. I will give him £200 for the slides and
library if you can arrange it with him – I can't pay for them until the
end of the year.[19]

Wright's wife and daughter, apparently in the mistaken belief
that Heron-Allen's intention was to keep the foraminifera in his
own collection, felt that they would rather sell to a museum than a
private collector. Accordingly in late August 1920 Swanston wrote
on their behalf to R.F. Scharff, who had been the acting director of
the whole of the National Museum since Count Plunkett had been

---

18 Keeper's Correspondence 1920/71, Cole to Harmer, 7 July 1920.
19 Keeper's Correspondence 1920/72, Heron-Allen to Harmer, 29 July 1920.

sacked in the aftermath of the Easter Rising:

> Mrs Wright and daughters, – with whom I sincerely concur, – would much prefer that the Collection should find a resting-place in an Institution available to the public, hence my addressing you with the view of you being able to secure it for your museum. And may I state, that, the Collection being largely derived from Irish localities it would, in my opinion, be very suitably and usefully housed in Dublin.[20]

After a short delay while Swanston had Wright's catalogue returned from South Kensington in order to send it off again to Dublin, Scharff approached T.P. Gill, still secretary of the whole Department of Agriculture and so in control of expenditure, to ask for a special grant of £200 to buy the collection. Gill asked for a reference for the grant, and by the end of October Scharff had got a suitable reference from G.H. Carpenter, the zoology professor at the Royal College of Science, insisting that the Wright collection was unique, and that it should be preserved for Ireland. Equipped with Carpenter's reference, Gill approved the grant, and Scharff wrote to Swanston on 1 November with the good news, enclosing pre-addressed labels to ensure speedy delivery.[21] Heron-Allen was naturally disappointed:

> Rather a blow! But it is my own fault; of course I ought to have given *you* the £200 to buy the Collection, & you could have 'lent' it to me for life, so that it would have been catalogued and ranged with all the other collections. But that's *that!*[22]

Cole and Scharff, who had both favoured bringing 'Irish' type specimens back to Ireland on previous occasions, were on opposite sides in this dispute. Cole was concerned for the availability of Wright's collection to future generations of researchers, but also for

---

20 Swanston to Scharff, 16 August 1920 (Natural History Museum Archives, Scharff correspondence).
21 Natural History Museum Archives, Official Correspondence drawer, Wright Collection folder.
22 Keeper's Correspondence 1920/72, Heron-Allen to Harmer, 11 November 1920.

the future of the relationship between the South Kensington Museum and Irish science. Scharff was responsible for building up his own museum's collections, and for safeguarding it as a national institution against whatever storms might lie ahead. He was able successfully to receive backing from T.P. Gill, who wanted to preserve the various parts of the department for which he had worked so hard for whatever form of government was to emerge. Cole, Heron-Allen and the British Museum do not seem to have even considered an equivalent political strategy on their side. At a time when British policy was explicitly for some form of Irish autonomy, it would probably not have succeeded even if they had.

The 'moving metropolis'

These three episodes are very reminiscent of the recently retold story of the Melbourne meteorites.[23] Two fragments of what was believed to be the largest known meteorite were identified near Melbourne in 1860. The ownership of the fragments was disputed between the National Museum of Victoria and the British Museum, represented respectively by the Irish palaeontologist Frederick McCoy (1823–99), the National Museum's director, and the German Ferdinand Mueller, the government botanist and keeper of the Botanic Gardens. Politicians in Victoria, including the colony's governor, became involved with the dispute, which ended with the larger fragment being moved to London and the smaller remaining in Victoria.

For Ireland's administration scientists, the location of the metropolis became distinctly uncertain between 1900 and 1922. Scharff would have had great difficulty even in initiating a dispute with South Kensington before the Dublin museum's administration was transferred to DATI in 1900. But once the Dublin Museum of Science and Art had become the National Museum of Ireland, Scharff could assert that it had a right to Ireland's national heritage, and that he was entitled to deny parts of that heritage to the imperial centre. Holt, also an Englishman working in the Irish scientific enterprise, could assert that specimens in Dublin 'would be useless' and

---

23 A.M. Lucas, P.J. Lucas, T.A. Darragh and S. Maroske, 'Colonial Pride and Metropolitan Expectations: The British Museum and Melbourne's Meteorites', *British Journal for the History of Science* 27 (1994), 65–87.

that South Kensington was 'the proper place for them' without needing to explain himself. Cole wanted to repatriate an important Irish fossil collection in 1915, but five years later did his best to prevent another collection being kept in Ireland. Whether Irish specimens belonged in a British or an Irish museum became a metaphor for Ireland's preferred political status among individual scientists. Scharff and Cole lost the disputes with Holt and the British geologists at times when Home Rule seemed a long way off; but Scharff was able to gain the Wright foraminifera when Irish independence was imminent, and the political balance shifting in favour of the National Museum of Ireland of which he was acting director.

12

State science in the
new state, 1920–1930

As has already been outlined in part I, the political situation in Ireland changed considerably between 1920 and 1922. The island was partitioned into two self-governing units by the Government of Ireland Act (1920), passed against a background of civil strife and the increasing ascendancy of militant Irish nationalism. Negotiations opened between the British government and Sinn Féin in the summer of 1921, concluding with the Anglo-Irish Treaty signed in the early hours of 6 December. Ireland was to become a Free State within the British Empire, though the six counties of Northern Ireland were given the power to opt out of the new arrangement and back into autonomy under Westminster. The treaty sparked a bitter dispute within Sinn Féin and the IRA, with a minority under the political leadership of Eamon de Valera refusing to accept the proposed relationship with the Empire, and the majority, led by Arthur Griffith and Michael Collins (both of whom were signatories of the treaty), prepared to work and administer the new system.

Once the political decision to transfer power had been made, the administrative transition took place in a matter of weeks. The functions of DATI in the six counties of Northern Ireland were transferred to the new government there on 1 January and 1 February 1922, while the Science and Art Institutions in Dublin, in common with most of the Irish administration of the twenty-six counties, were transferred to the Provisional Government, chaired by

Michael Collins, on 16 January after the Southern Irish parliament had ratified the treaty.[1] The Fisheries branch of DATI became a separate ministry. Two days later Collins appointed Patrick Hogan (1891–1936) as the first Irish minister for agriculture, a post he retained until his party lost office in 1932. Hogan was a solicitor and small farmer whose robust advocacy of free trade occasionally embarrassed his more protectionist colleagues.[2]

The Science and Art Institutions stayed as a single administrative unit until 1926, but remained under Hogan's control only until 1924 when they were transferred to the department of education by the provisions of the Ministers and Secretaries Act. The minister for education appointed in 1922 was Eoin MacNeill, professor of early and medieval Irish history in UCD, whose troubles with the Royal Irish Academy were recounted in part I. MacNeill was also the Irish Free State's nominee on the ill-fated Boundary Commission, and resigned from public office when it collapsed in 1925. Another UCD historian, John M. O'Sullivan, who held the chair of modern history in the college, replaced him until the 1932 election.[3]

During the 1914–18 period, the state scientific institutions had lost both resources and personnel to the war effort. Reconstruction after the end of the war had been hampered by uncertainty – what government would be running the new institutions? Vacancies caused by the war had not been filled, but the new government came into office pledging to develop Ireland's resources to compensate for centuries of supposed British neglect. The political instabilities of the years leading up to 1920 had made it difficult for any administration scientist to plan ahead for the new era. This chapter investigates the reasons for the departure of one of Ireland's younger geologists and Cole's failure to acquire qualified staff for the Geological Survey. It also describes the attempts by the Royal College of Science to gain some security through a closer association with Trinity College, and

1   J. McColgan, *British Policy*, pp. 69 and 73 (n. 76).
2   By an unusual provision of the Free State constitution, between December 1922 and 1927 Hogan remained a minister but was not a member of the cabinet. See Donal O'Sullivan, *The Irish Free State and its Senate: A Study in Contemporary Politics* (1940), p. 184.
3   See John Coolahan, *Irish Education: Is History and Structure* (1981), for a general review of Irish education ministers and their policies since independence.

the collapse of that scheme when the college was closed during the civil war and then annexed by University College Dublin. It concludes by briefly looking at the development of the Natural History Museum, and by examining the models which have been put forward to explain the role of science in independent Ireland.

Departures from geology

Many Irish Protestants in the twenty-six counties had reconciled themselves before 1921 to an accommodation with nationalism.[4] Those employed in the civil service who were unable to accept the peace settlement were protected by a clause in the treaty which allowed them to resign with full pension rights if they did not wish to serve under the new government. Several of the state's scientific employees departed thus. These included G.H. Carpenter, the professor of zoology in the Royal College of Science, who became the keeper of the University Museum of Manchester and eventually took Holy Orders in the Anglican church. One administration scientist who had already left Ireland, even before the treaty was signed, was the geologist W.B. Wright, who also went to Manchester to organise a branch office of the Geological Survey in Great Britain.[5]

There are two versions of the reasons for Wright's departure. Oral tradition in the Geological Survey had it that he 'was the victim of some "incident" and in consequence he in March 1921 took a transfer to a Survey post in England'.[6] It has not been possible to discover any more about this 'incident', although early 1921 was much the bloodiest period of the 'War of Independence' and perhaps one does not have to look further than the generally violent environment of the time. On the other hand, Darrell Figgis, an independent TD for

---

4   David Fitzpatrick, *Politics and Irish Life, 1913–1921: Provincial Experience of War and Revolution* (1977), provides a penetrating insight into shifting attitudes in Co. Clare.

5   As far as I am aware, W.B. Wright was not related to Joseph Wright of the foraminifera. His obituary is in *Nature* 144 (1939), 775–6. Herries Davies recounts the decline of the Geological Survey in the 1920s and 1930 in ch. 6 of *North from the Hook* ('Paralysis 1924–1952', pp. 153–74).

6   Herries Davies, 'Irish Thought in Science', in *The Irish Mind* (1985), p. 308. However, in *North from the Hook*, p. 231–2, Herries Davies admits that there is no evidence apart from oral tradition to support this account.

Co. Dublin, speaking to the Dáil on 25 June 1923 seems to have interpreted Wright's departure as a purely financial decision (admittedly Figgis was speaking to the estimates for the Science and Art Institutions, and so would naturally concentrate on the financial aspect):

> Last year the Senior Geologist of Ireland, who is now no longer there, was offered twice the salary he was receiving in Ireland from one county in England to undertake its mere county survey, and he naturally left. That is not a very happy, a very desirable, nor a very hopeful state of affairs.[7]

Wright was supporting his elder brother and bedridden mother as well as his own family, so the financial incentive to move is clear. His daughter, who was born in 1917, has no memories of any 'incident' involving her father:

> In 1914 my father published his magnum opus *The Quaternary Ice Age*. This gained him immediate recognition in England as well as Ireland and indeed it remained a standard work for both geologists and geographers until after the Second World War and is still a recognised classic. He must have felt that his opportunities would be greater in England once the Irish Geological Survey separated from its parent body . . . To these possible positive reasons for leaving Dublin may have been added a fear of what would happen under the new regime in Dublin. He was after all in a very secure position as a British scientific civil servant.
>
> [By leaving Ireland for Manchester] he also retained his British nationality and this could have been a factor though I doubt if it would have been more than marginally significant. My family belonged to the protestant minority in Ireland and they may have had some fears for their future. In the event the new government treated the protestant minority very well and that part of my family who remained in Dublin have become very loyal citizens of Eire.[8]

---

7   *Dáil Debates* III, 2405. Figgis implies that Wright left in 1922, but all other sources are in agreement that it was 1921. It is a considerable exaggeration to say that the Manchester post had 'twice the salary'; according to Herries Davies (*North from the Hook*, p. 232) it was £650 rather than £600.

8   Letter to NHW from Barbara, Lady Dainton, 27 January 1992.

Wright's move and the effects of the war casualties, still not replaced, left the Geological Survey with three vacancies out of five scientific posts. As Wright's departure drew near, Cole offered a survey position to C.A. Matley (1866–1947), a recently retired Indian civil servant who had done some geological work in Ireland fifteen years before and was now looking for a geological post. Matley's initial response was unenthusiastic, but Cole replied defensively that Ireland was not as bad as it was portrayed:

> As you remark, Dublin is not a pleasant place to settle in just now – no part of Ireland is at present; but it is wonderful how little one's ordinary avocations are interfered with, and scientific work keeps one out of many complications.[9]

Matley however was not to be convinced, and indeed shortly afterwards became the government geologist of Jamaica, where we may assume that he found the climate more congenial:

> [M]y wife does not view with favour any proposal to reside and work in Ireland during the present unsettled state of the country. As you know, a field geologist is looked upon in most parts of the world as either a harmless lunatic or as having some ulterior motive, and a man wandering around the Dublin mountains (a reputed drilling ground of the Sinn Féiners) with a map and a notebook is likely to be regarded by both sides with suspicion; & the recent attitude seems to be to shoot first & enquire afterwards.[10]

The Geological Survey's scientific posts were all filled at the end of 1921, but not in a very satisfactory way. Hallissy became senior geologist, and the three posts of geologist were filled by John de Witt Hinch (1875–1931), a former employee of the National Library who had been the survey's superintendent of maps since 1919, and by two new recruits, T.J. Duffy and Anthony Farrington. After Cole's death in 1924, Hallissy had to wait for four years to be promoted to fill his place (although this was because of the ambiguous status of the chair of geology in the Royal College of Science which Cole had

---

9   Cole to Matley, 10 February 1921, in GSI file 11B.
10   Matley to Cole, 15 February 1921, in GSI file 11B. See also Herries Davies, *North from the Hook*, p. 150.

also held). Hinch's old post was not filled, and Duffy and Farrington were employed on only temporary contracts for seven years. The survey was transferred from the Department of Education to that of Industry and Commerce in 1927. Farrington resigned in 1928 to spend the rest of his working life at the Royal Irish Academy, where he was able to contribute considerably more to Irish geology than he might have done had he stayed with the survey, and Duffy was finally given a permanent contract in 1929.

The closure of the Royal College of Science

The Irish civil war began on 28 June 1922 when forces loyal to the Collins government attacked a garrison of Irregular troops, opposed to the treaty of the previous December, which had been occupying Dublin's Four Courts. The active phase of the war lasted hardly two months, but by then the two leading figures on the pro-treaty side were dead, Arthur Griffith of a stroke on 12 August and Michael Collins in an ambush ten days later. The incipient government was taken over by W.T. Cosgrave, who was to remain in power for the next ten years. Kevin O'Higgins, the Minister for Home Affairs, vividly described the new Cabinet as:

> . . . simply eight young men. . . standing amidst the ruins of one administration, with the foundations of another not yet laid, and with wild men screaming through the keyhole. No police was functioning through the country, no system of justice was operating, the wheels of the administration hung idle, battered out of recognition by the clash of rival jurisdictions.[11]

The Third Dáil convened on 9 September 1922, and for the first time, Dáil Éireann was recognised internationally as the Irish parliament. The story of the government's negotiations with the RDS, resulting in the permanent occupation of Leinster House by the Dail, has been told in part I. The National Library and National Museum were closed to the public for the duration of the civil war. The new government's ministers, still under threat of assassination by the

---

11 Terence de Vere White, *Kevin O'Higgins* (1948), p. 83, quoting from a speech made by O'Higgins in 1924.

increasingly desperate Irregulars, became uncomfortably aware that the Royal College of Science, which shared a building with most of the civil service machinery of administration and backed onto Leinster House where the Dáil was meeting, was a weak point in their defences – and once the teaching term began, its student body would potentially have access to all kinds of dangerous materials in the workshops.

The college had experienced a considerable boom since 1918, as a large number of students whose courses had been interrupted by the war returned to complete their studies. The published figures for the years up to 1924 are not as complete as in earlier years, but they illustrate the situation well enough (see Table 12.1).

Table 12.1: Students at the Royal College of Science for Ireland, 1917–1924

Academic Year	Associate	Non-Associate	Associate students by year of study			
			1st	2nd	3rd	4th
1917–1918	98	96	(not available)			
1918–1919	127	131	(not available)			
1919–1920	214	217	(not available)			
1920–1921	240	131	(not available)			
1922–1923	222	67	34	39	66	83
1923–1924	160	16	25	32	51	52

Source: DATI and (Royal) College of Science for Ireland annual reports.

The Royal College of Science staff were probably not as concerned about the potential security implications of their building's geographical position. During the late spring and summer of 1922, however, an attempt had been made to safeguard the college's administrative position by seeking a closer alignment with Trinity College. The provost of Trinity, J.H. Bernard, made an approach to T.P. Gill on 7 April seeking closer co-ordination between the two institutions, in particular with regard to their agriculture courses.[12] Bernard was rather preoccupied over the next few months as one of the representative Southern Unionists negotiating safeguards for the Protestant minority in the new state with the British and Irish governments, but over May and June a joint committee of Trinity and the Royal

---

12   Trinity College Dublin (J.H. Bernard papers), MSS 2388–93/328, and subsequent correspondence, esp. /347, /348, /363.

College of Science agreed that they would gradually move to a position where Trinity would award degrees to students of the Royal College. Gill was clearly aware that this was potentially a political rather than an administrative decision, and was worried enough to write to Bernard from London on 20 September with advice on how the matter should be handled:

Dear Provost,

My Secretary has sent me on your letter of the 12th inst., (I am at present trying to take a short holiday at this side).

A note of the further facilities is being incorporated in the College of Science Calender [*sic*]. We are treating the matter as a more natural extension of the co-ordination *already in existence* – which is exactly what it is. This is the wisest course at the moment. Any advertisement or calling special attention to the matter would be inadvisable in the abnormal state of things we are passing through and might lead to a misunderstanding. I venture to recommend that you adopt the same course. Just put it in your calender [*sic*] as the College of Science are doing in theirs. I am asking them to send you a copy.[13]

The advantages on both sides were clear. Trinity, as has already been stated, was keenly aware that it had missed out on the investment promised by the 1920 Royal Commission, and was looking for ways to make itself essential to the state apparatus. The Royal College of Science had always had the disadvantage of being unable to offer its students full degrees, and perhaps also hoped that its association with the much older institution might protect it against arbitrary changes implemented by the government.

But these carefully laid plans were to come to nothing. On 27 September the Provisional Government issued an order to the college's staff through DATI's assistant secretaries, Fletcher and Campbell, prohibiting all access to the building until further notice and directing the staff to carry on as much work as possible in such additional premises as could be secured elsewhere.[14] Although

---

13   TCD MSS 2388–93/364.

14   'Calendar of Events Leading up to and Arising out of the Closing of the College Building, Merrion Street, Dublin', in the minutes of the College's Council Meeting, 13 December 1922 (UCD Archives RCSI/11); hereafter 'Calendar of Events'.

there was no longer any house-to-house fighting in Dublin of the kind there had been in the first week of July, sporadic shootings and bombs continued in the capital and several major towns were still under Irregular control, so the security precautions could certainly be justified. The College of Science in the meantime negotiated the use for lectures of the old UCD building at 86 St Stephen's Green through MacNeill, the Minister for Education. First-year students were to be admitted to the equivalent UCD classes. Laboratory facilities were found in the Albert Agricultural College, Botanic Gardens, Veterinary College, Geological Survey, and the Annexe which was all that remained of the 1907 Industrial Exhibition. This still left no suitable facilities for chemistry, physics or engineering subjects.[15]

On Monday 9 October classes began in the dispersed college. Some equipment had been evacuated from the Merrion Street buildings, and the staff were resigned to the possibility that they would not return to the college buildings until after Christmas, when the civil war might have settled down. The students were less happy with this state of affairs, and on 13 October the students' union sent a deputation to the minister of agriculture, Patrick Hogan, to request the re-opening of the college as soon as possible. The newspapers reported this the next day, along with the unofficial government explanation that there had been a plot to blow up the college, with inevitably catastrophic consequences for Leinster House where the Dáil was meeting and for the Provisional Government offices in the northern wing of the College building.[16] There is no reference to any such plot in any history of the civil war. During the summer of 1922 some Irregulars had planned to steal an aeroplane and drop bombs from it onto the Leinster House area, but the project was abandoned.[17]

Perhaps the students should not have reminded the minister about the situation. Two days later, on Sunday 15 October, he resolved to take direct action about the lack of laboratory facilities available to

15 'Report on the Arrangements for the Work of the College, Session 1922–23' (UCD Archives RCSI/62); hereafter 'Report on Arrangements'.
16 *Irish Times* and *Freeman's Journal*, 14 October 1922.
17 See C.S. Andrews, *Dublin Made Me: An Autobiography* (1979), pp. 237–9, for details.

the college's students. According to Fletcher, who relayed his message to the stunned college staff, Hogan was not satisfied that the college was using the laboratory facilities of UCD to the fullest possible extent. The College of Science[18] interpreted this to mean that the minister wanted *all* college activities to be transferred to UCD as soon as possible. This was confirmed on Monday 16 October, when another memorandum arrived via Fletcher directing first-year students to report to UCD on Tuesday (the next day), second-years on Wednesday, third-years on Thursday and fourth-years on Friday. The college's staff and students both held angry meetings to decide their course of action. The unhappy Fletcher, whose task it had been to pass on the minister's wishes, found time to write to Gill, who was in London:

My Dear Gill,

Matters develope [*sic*] rapidly. The Minister for Agriculture has instructed me that all the work of the College of Science, as far as possible, is to be done in the Univ College (Earlsfort Terrace) and has appointed Prof Drew [Royal College of Science professor of Agriculture] to act as intermediary between me and Dr Coffey [UCD president] whom I have seen several times with Heads of Faculties. We had already made very satisfactory arrangements – under the circs – for the work to be done at 86 St Stephens Green (old U.C.) Albert College Vet. College, [Industrial] Annexe and Geological Survey. Univ. College (through Drew) undertakes *all* the work and the First Year Students are being instructed to go to Univ College tomorrow. The others will probably follow on Wednesday & Thursday. This will leave our Staff without work for the time being. I am proceeding as far as possible in conjunction with Campbell & the College Council – we met them this afternoon. I write you as soon as possible for it is only today that I was told to definitely concentrate the work in Univ. College. Meantime one or more Ministries are securing temporary accommodation in the College

---

18 The use of the word 'Royal' in the college's title remained a matter of some confusion. The college's own records use it consistently; the *Dáil Debates* and *Freeman's Journal* as consistently omit it. The *Irish Times* appears to use the word 'Royal' for the first appearance of the College in the body of an article and to omit it thereafter, rather as the BBC are supposed to deal with a well-known city and county in Northern Ireland today.

Buildings – though I have only been informed through Quane [the Royal College's registrar] – I am expressing no opinion on these changes for my opinion is not asked – I am merely carrying out an instruction.[19]

The closure of the College of Science had been transformed from a minor administrative inconvenience to a major political embarrassment for the new government. That afternoon the students' union resolved by seventy-nine votes to forty-two 'that the students of the College of Science refuse to go to the National University'.[20] The meeting was fully covered in the next day's papers, and a sympathetic editorial in the formerly Unionist *Irish Times* voiced the suspicions that had perhaps been circulating for some time:

Rightly or wrongly, a strong impression prevails – it has been conveyed to us from different quarters which are thoroughly loyal to the Free State – that the Government's decision spells the death-sentence of the College of Science. Already some parts of the great building in Upper Merrion street are being used as Government offices, and the talk is that Science will never return to her noble abode – that the Government, through motives of economy, has turned her adrift, leaving her to knock at the door of universities which, however hospitable, cannot give her what she needs.[21]

The teaching staff had by this time contacted Professors William Magennis and William Thrift, who were respectively members of the Dáil for the National University and the University of Dublin (i.e. Trinity College), to lobby the minister on their behalf. Their grievance, as it was later expressed, was that the government had unjustly and unnecessarily interfered in the rights and privileges of the council of the college as an academic body, and that the arrangements that had been made for the students before the minister's personal intervention were perfectly adequate. The staff drew up a four-point memorandum to be presented to the minister, which included a potential development plan for the future of the college

---

19 National Library of Ireland (Gill papers), MS 13490/15, 16 October 1922.
20 *Irish Times* and *Freeman's Journal*, 17 October 1922.
21 *Irish Times* editorial, 17 October 1922.

once they regained access to the buildings.[22] The students meantime held a public meeting 'to discuss what steps should be taken in connection with the taking over of the College buildings in Merrion street by the Irish Provisional Government for use as offices':

> The Chairman [of the Students' Union] said they had been informed that it was a military necessity that the College should be taken over by the Government. He had been informed that there were bombs and mines found there, but he had it now on good authority that there was absolutely nothing found in the College, and that did away to a great extent with the plea of military necessity.[23]

> Another speaker declared that the bomb which, it was reported, had been found in the College was simply an aeroplane dynamo for demonstration purposes.[24]

The traditionally nationalist *Freeman's Journal* editorial the next day (19 October) warned against stirring up anti-government sentiment during a time of political crisis:

> The disturbance of the work of the College of Science is to be very much regretted, but the professors and students will make a mistake if they try to turn the quite natural movement for the protection of their interests into a semi-political agitation. The Minister for Defence[25] is himself a student of science and would be the last man in

22 'Calendar of Events', p. 7.
23 *Freeman's Journal*, 18 October 1922.
24 *Irish Times*, 18 October 1922.
25 Richard Mulcahy (1886–1971), originally a post-office engineer, had in fact been awarded a scholarship by the Royal College in 1911 but had been unable to get permission for the three years' leave of absence that would have been necessary to take a full-time course. He was briefly Minister of Defence in the first revolutionary Dáil government in 1919 and was appointed to that post again by Collins in the January 1922 Provisional Government. From late 1919 he had also served as Chief of Staff of the IRA and then of the Pro-Treaty forces, taking over as Commander-in-Chief after Collins was killed in August 1922. The potential conflict between his two roles as a member of the government and a representative of the army led to his resignation from both positions in the wake of the 'Mutiny' of March 1924, although he served in three subsequent cabinets and was the leader of Fine Gael from 1943 to 1959. See Maryann Giulanella Valiulis, *Portrait of a Revolutionary: General Richard Mulcahy and the Founding of the Irish Free State* (1992). Roy Foster's biographical note on Mulcahy in *Modern Ireland*, p. 511, contains several errors, notably that he was a post office clerk rather than an engineer.

Ireland to disturb a home of science unless compelled by a stern necessity. By the stupid arrangements of the Board of Works, the college and the principal government offices came to be housed practically under the same roof, and in the present condition of affairs that arrangement has been found to be unworkable.[26]

The Minister for Agriculture, questioned that day in the Dáil by Thomas O'Connell, a Labour TD from Galway, defended the closure of the college on the grounds of (unspecified) military necessity and pointed out that facilities had been provided by UCD and were on offer from Trinity. However, he went on to hint at the possibility that the arrangement between the College of Science staff and UCD could develop into a formal merger between two institutions:

> There are a great many anomalies connected with the College of Science. It is an anomaly to have one educational establishment in Merrion Street under the Ministry of Agriculture, and another at Earlsfort Terrace [UCD] under the Minister of Education, both teaching to a great extent the same subjects. There is no reason whatever for duplication so far as there is duplication, and this question will have to be considered when the College of Science is reconstructed.[27]

The next day O'Connell, with support from his party leader, Thomas Johnson,[28] and the Trinity TDs Sir James Craig,[29] William Thrift and Gerald FitzGibbon raised the closure of the college on the adjournment debate.[30] If there was a real security risk, why had the houses on the far side of Merrion Street not been closed? Was the real reason for the closure of the college in fact that the government needed office space? Was it true, as the students' union had alleged earlier that day, that the Board of Works had prepared plans for the

26 *Freeman's Journal* editorial, 18 October 1922.

27 *Dáil Debates* I, 1654 (18 October 1922).

28 Thomas Johnson (1872–1963), leader of the Irish Labour Party from 1922 to 1927, not to be confused with his namesake, the professor of botany at the Royal College of Science.

29 This was the professor of veterinary medicine at Trinity College, not the future Lord Craigavon, the first prime minister of Northern Ireland, whose name was also Sir James Craig.

30 *Dáil Debates* I, 1799–1814 (19 October 1922).

internal reconstruction of the college which had already been started? What precisely did the minister mean by 'reconstruction'? The unspoken question seemed to be, was the government in fact planning to take over the college buildings for its own purposes while merging the college by fiat with UCD?

Hogan reiterated that military considerations alone had been behind the closure of the college buildings; that the best possible provision for the students was being made with UCD; that there would have to be a 'reconstruction' of the college because it was 'abnormal' to have two similar educational institutions under two different ministers; but that the government would certainly use rooms in the building as office space if the need arose. MacNeill, the minister for education, disclosed that Hogan had been ill at the time the decision to close the college had been made and in fact had not been consulted. There were no 'ulterior designs' behind the closure:

> With regard to the future of the College, there is not the slightest reason to suppose that the Government does not desire that the benefits of the College of Science to the public shall be continued, but the Government certainly does not pledge itself that they would be continued in any particular form . . . The Government, so far as it can, will maintain the benefits of this College for the students who attend it, and that is all they pledge themselves to do.[31]

Earlier that day (or possibly the previous day)[32] Hogan had in fact come to a settlement with the college staff as a result of the mediation of Magennis and Thrift. The corporate entity of the Royal College of Science was to be continued; the college's council would retain academic authority; the specialised equipment in the college buildings would not be altered without permission from the Dean. In return for this recognition, the staff undertook to make the best of the facilities that had been provided in other institutions and to recommend to the students that they do the same. Those students who wished to transfer to other universities could do so. The agreement was put to a meeting of the students' union the following day

31 *Dáil Debates* I, 1813–1814.
32 The 'Calendar of Events' is unclear.

(20 October) and somewhat begrudgingly endorsed by seventy-one votes to eight with a number of abstentions.[33]

There matters remained for some months, although resolutions urging the government to re-open the college as soon as possible were passed by the Royal Dublin Society's Committee of Science and its Industrial Applications (14 November),[34] the board of Trinity College (21 November),[35] and the Free State's own senate (8 February 1923).[36] After Christmas, several fourth-year students who had been studying applied chemistry and engineering successfully pleaded that while the college remained closed, the alternative laboratory facilities provided were inadequate, and they were allowed to transfer to the University of Birmingham and University College London respectively – all were ex-servicemen on grants from the British government.[37] The engineering lecture rooms were re-opened shortly after, and finally in June third-year electrical engineering students were permitted to use their laboratories – as long as they managed to obtain a guarantee of their loyalty to the Free State from a senator, TD or similar person.[38]

The Dáil debated the ultimate fate of the college at great length in July 1924.[39] MacNeill, who was by now responsible for the Science and Art Institutions, revealed that the government did indeed plan to merge the entire College of Science with UCD. The government did not wish to be directly responsible for the administration of a teaching institution, while UCD had suffered from historic underfunding and was not well equipped for science or in particular engineering. This met with understandable though polite

---

33 *Irish Times* and *Freeman's Journal*, 21 October 1922.

34 Royal Dublin Society Archives, minute book of the Committee for Science and its Industrial Applications, pp. 155–6. George Fletcher, the assistant secretary of DATI who had been Hogan's initial emissary to the college staff, chaired this meeting.

35 'Calendar of Events'.

36 *Senate Debates* I, 291–8.

37 Darrell Figgis raised this in the Dáil on 17 January (*Dáil Debates* II, 843).

38 This paragraph summarises the second page of the typed 'Report on Arrangements', and is also in agreement with the published annual report for 1922–3 (in the appendix to *23rd DATI Report, 1922–3*, p. 196).

39 *Dáil Debates* VIII, 1047–105 (11 July), 1607–30 (17 July), 1683–715 (18 July) & 1727–58 (21 July).

resistance from the TDs for Trinity College, who received some support from Johnson, the Labour leader, and from both representatives of the tiny Businessmen's Party; but they were faced with a *fait accompli*. The College of Science was wound up as an institution, and its staff and buildings transferred to UCD, by the University Education (Agriculture and Dairy Science) Act of 1926.

The college buildings remained in the control of the Engineering School of UCD until 1989, when, as the students' union had feared would happen in 1922, the entire building became government offices. The UCD staff joined the rest of their college in a new building on the Belfield site to the south of the city. An allegory on the college and its neighbouring government departments, printed in a College of Science student magazine in 1916, had ended with the prediction that 'the Rattlesnake eats the Owl and compels the Prairie Dog to look for new digs'.[40] We too should refrain from allocating the titles, and shall merely note that the present occupant of the old Royal College of Science building is the Office of the Taoiseach.

A decline in Irish state science?

The story of the National Museum during this period is less dramatic. It remained closed during the 1922–3 civil war. When it re-opened there had been considerable changes to its layout and staff. Scharff had retired in 1921, and was replaced as keeper of natural history by A.R. Nichols. The vacancy created by Selbie's death in 1916 was filled in 1920 by the appointment of A.W. Stelfox (1883–1972), an experienced naturalist from Belfast. J.R. Halbert retired in 1923, and Nichols in 1924. Their places were not filled, and all the natural history work of the museum now devolved upon the irascible Stelfox, Matilda Knowles and Eugene O'Mahony, a technical assistant appointed in 1922. The Dáil and senate had been sitting in Leinster House since September 1922, and this had caused considerable disruption to the available exhibition space – the curved corridor where the geological collection had been displayed was taken over for office purposes, and the increasingly dilapidated Natural History Annexe was closed. Plans to rebuild it in 1931 fell

---

40 *The Torch*, vol. 1, no. 1, January 1916, p. 2 (surviving copy in the National Library of Ireland).

through due to a personality clash between Stelfox and the newly appointed keeper of natural history, Patrick O'Connor. It was demolished in 1961 to make way for a restaurant for TDs and senators.

A Committee of Enquiry was set up by Prof J.M. O'Sullivan, the Minister for Education, in 1927 to report on the museum as a whole, with as its special adviser Nils Lithberg of the Northern Museum in Stockholm. Its report recommended that the museum concentrate on Irish antiquities in general (incidentally bearing out Sheets-Pyenson's model of a shift to local studies in the period of increasing autonomy), but also recommended that the natural history staff be increased to a keeper, assistant keeper, four assistants and two technical assistants.[41] In fact, the established professional staff on the natural history side were reduced in number from five to two. The only outstanding vacancy to be filled was the post of keeper.[42]

It is possible to interpret the slowing of government expenditure on these scientific institutions as a symptom of a general antipathy between Irish nationalism and the scientific enterprise. Roy Johnston certainly seems to interpret the history of Irish science in general in this light. In an essay on the British Association meetings held in Ireland, he comments *apropos* de Valera's participation in the 1908 meeting:

> Apart from him, one looks in vain for any evidence on the part of the (again rising) national movement for appreciation of the importance of science and technology in the nation-building process.[43]

But the case against Irish nationalism as a despoiler of Ireland's scientific talents is not proven by the events described here. In the years after 1900, there was a substantial investment in Irish state science by British governments, whether or not they favoured Home

---

41 'National Museum: a) Report of Committee of Enquiry, 1927; b) Report of Dr Lithberg, 1927' (duplicated typescript; copies exist in National Library of Ireland and Trinity College Dublin).

42 O'Riordan states that Lithberg suggested this reduction. However, Lithberg's written suggestions for the Natural History wing of the museum differed from the recommendations of the committee as a whole only by the omission of the new post of assistant keeper.

43 Roy Johnston, 'Science and Technology in Irish National Culture', *Crane Bag* 7 (1983), 58–63.

Rule. English scientists working in Ireland in the years before independence certainly did not necessarily see their interests and the interests of nationalism as divergent. After independence, the work of the College of Science continued, although in a new form; the Geological Survey continued to survey and produce occasional memoirs; the Ministry of Fisheries continued to carry out coastal research, if not quite on the scale of the early years. Only in the National Museum was science directly forced to make way for the instruments and symbols of nationhood, in the shape of the Dáil's physical presence and an increased emphasis on Irish antiquities and revolutionary souvenirs among the exhibitions. But the Cosgrave government was not hostile to modernity *per se*. If it had been, it would hardly have funded the Shannon hydroelectric scheme.

The common thread is not the downgrading of science but the cutting of government expenditure. The College of Science was disposed of to UCD, rather than spending more money to build UCD its own science and engineering facilities. Vacancies were left unfilled and promotions delayed. The fiscal conservatism of the new government, or rather of the two chief officials at the Department of Finance, Joseph Brennan and J.J. McElligott, is legendary. Lee points out that the entire Irish civil service in 1932 cost £3.9 million per year for over 23,000 employees, compared with £4.2 million for 21,000 ten years earlier.[44] The new Free State was not hostile to modernity. It was simply hostile to spending money. We will explore more fully the relations between the various strands of Irish nationalism and of the scientific endeavour in part III.

---

44 See J.J. Lee, *Ireland 1912–1985: Politics and Society* (1989), p. 108.

Part III

13

Nationalist scientists

Three different surveys, quoted at the start of an earlier chapter, found that the majority of nineteenth-century Irish scientists had Ascendancy backgrounds and that the estimated proportion of Catholics is somewhere around 10 or 15 per cent.[1] This is not negligible, but it is well below the 70 to 75 per cent who were Catholics. Some writers have suggested that either the Roman Catholic church, or Irish nationalism, or perhaps both, were in some way responsible for inhibiting the growth of science among their respective followers before 1921 and for stunting it afterwards. So for instance John Wilson Foster writes of 'the calculated exclusion of science, particularly natural science, by the architects of the Irish Cultural Revival around the turn of the century'.[2] Roy Johnston blames religion rather than politics:

> Cardinal Cullen blocked Catholics from going to the 'Godless Colleges', and access to scientific technology for the sons of the rising Catholic bourgeoisie was severely restricted for the next 50 years, until the NUI was set up . . . Protestant predominance in 19th century Irish science must be attributed to Cardinal Cullen. [Nicholas] Callan showed what the potential was, and it was strangled at birth.[3]

---

1 See the introduction to part I above.
2 John Wilson Foster, 'Natural Science and Irish Culture', *Éire-Ireland*, 26, no. 2 (1991), p. 95.
3 Johnston, 'Godless Colleges and Non-Persons', *Causeway* 1, no. 1 (September 1993), pp. 36–8, p. 38.

We have seen that the picture is considerably more complex. The extent to which discrimination against Catholics existed in the institutions of the Ascendancy has been demonstrated in part I. While Johnston is right to point out that a crucial factor in discouraging Catholics from following scientific careers was the lack of an education system which they were prepared to buy into, it is clear that the responsibility for this state of affairs was more widely shared than he allows. If there had indeed been systematic opposition to the study of science *per se* from the Catholic hierarchy, the evidence of this could hardly have been suppressed. In fact the record shows that the opposite is the case.

Because there were not many nationalist/Catholic scientists in the first place, and because even they are under-represented in the historiography of Irish science, there has been a tendency to see the situation as more one-sided than it in fact was. While published sources on Trinity College, the RDS and the Science and Art Institutions are plentiful, only one of the three constituent colleges of the National University has recently inspired an institutional history. Primary as well as secondary material is scarce. Nationalist scientists, and those charged with the task of administering them, seem to have left fewer written records than have those working in the institutions of the Ascendancy or the administration, and since this book originated in a search for extant manuscript resources it will reflect this fact. There is certainly plenty left to say about the nationalist tradition within Irish science, which has been neglected by historians both of Irish politics and of Irish science.

14

Catholics in Irish science

The low proportion of Catholics among Irish scientists of the nineteenth century cannot be denied. One of the reasons for this shortfall suggested by a number of writers is that the Catholic church itself is intrinsically inimical to the development of science. The normally perceptive John Wilson Foster, for instance, writing about the low proportion of Catholics among the members of the nineteenth-century Belfast Naturalists Field Club, suggests that perhaps 'the uneasy relationship between the Catholic church and the study of Nature was a factor' as if the 'uneasy relationship' were a well-established fact. This assumption will be challenged here.[1]

If there had indeed been a policy of discouragement of scientific research by the Catholic church, we would expect to find that scientists had been regularly censured by ecclesiastical authorities for their research, and discussion of scientific topics banned from ecclesiastical journals. In fact we will find that theological colleges supported their staff's scientific research and plenty of scientific discussion in the Irish Catholic journals. It is certainly true that

---

1 J.W. Foster, 'Natural History, Science and Irish Culture', *Irish Review* 9 (1990), 61–9, p. 63; see also his 'Natural Science and Irish Culture', *Éire-Ireland* 26, no. 2 (1991), 92–103, and 'Natural History in Modern Irish Culture', in *Science and Society in Ireland*. Greta Jones, 'Science, Catholicism and Nationalism', *Irish Review* 20 (1997), p. 47–61, also criticises the assumption that nationalism and Catholicism are inimical to science, though she finds less reason to defend the Catholic church than I do.

the church had a particular ideological axe to grind as far as science went; but it is increasingly apparent that this was just as true of other institutions in Irish society as well.

The Merton thesis

Support for the idea that Rome and science don't mix can be found, for example, in one of the most influential studies of the relations between religion and the history of science, Robert Merton's 'Science, Technology and Society in Seventeenth Century England', often referred to as the 'Merton thesis'. In summary, Merton argued that 'Puritanism, and ascetic Protestantism generally, [were] an emotionally consistent set of beliefs, sentiments and action which played no small part in arousing a sustained interest in science'. Although the main thrust of Merton's argument concerned the relationship between English Puritanism and science in the scientific revolution, he also considered the situation in other countries and at later times, up to the end of the nineteenth century.[2]

Merton believed that a fundamental difference between Catholic and scientific mind-sets persisted. He showed that Protestants were over-represented and Catholics under-represented in the more mathematically oriented *Realschulen* in three different German states at the end of the nineteenth century. Comparing Protestant Scotland and Catholic Ireland, he quoted Havelock Ellis's *A Study of British Genius* (1904):

> In science Scotland stands very high, Ireland very low . . . In order to realise the extraordinary preponderance of the Scotch over the Irish contingent, it must be realised that until the present century the population of Ireland has been much larger than that of Scotland, and it may be noted that the one purely Irish man of science (Tyndall) was of original English origin.[3]

Merton went on to assert that the Irish excelled in the histrionic arts, making up for their lack of science. This sort of sweeping

---

2   Robert Merton, 'Science, Technology and Society in Seventeenth Century England', *Osiris* 4 (1938), 360–632, esp. pp. 486–95.

3   Ellis, *A Study of British Genius* (1904), pp. 66–7; Tyndall would perhaps not have appreciated being described either as 'purely Irish' or 'of original English origin'.

generalisation, with its roots in eugenics and anti-Irish racism, has no place in serious historical analysis today. Ellis's theories on genius were refuted by Jacques Barzun as long ago as 1938.[4] More recently, D.S.L. Cardwell has dismissed Merton's case:

> The Irish, Merton comments, make up for their lack of science by excelling in the histrionic arts, which are disapproved [of] by Calvinists. But an Irishman could reply, very fairly, that Calvinist Wales has yet to produce one world-famous scientist while it is undeniable that the Welsh are successful practitioners of the histrionic arts, however repugnant they are supposed to be to Calvinists.[5]

The main part of Merton's thesis, that Protestantism and science in seventeenth-century England are linked, has been the foundation for much subsequent study of the Scientific Revolution. However, its relevance to the social structure of science two centuries later is questionable.

A more helpful analysis of the lack of Catholic scientists in Ireland must take into account both the relative poverty of Catholics compared to Protestants (with the consequent relative lack of educational opportunity) and the fact that Ireland was a divided society at the time. Herries Davies misses the latter problem in his 1985 essay on 'Irish Thought in Science',[6] in which he compares the significant number of English and Scots scientists in the nineteenth century of 'humble origins' with the low number of Irish equivalents. We have already examined the barriers which applied to Catholics in the specific cases of Trinity College, the Royal Dublin Society and the Royal Irish Academy. It should be emphasised again that these barriers were symptoms of a society where the Ascendancy remained ascendant until the turn of the century.

Science and the Irish church

The evidence for any intervention by the Catholic church intended to deter scientific research in the case of Ireland is almost non-existent. It

---

4   Barzun, *Race: A Study in Modern Superstition* (1938).
5   D.S.L. Cardwell, *The Organisation of Science in England* (1972).
6   G. Herries Davies, 'Irish Thought in Science', in *The Irish Mind*, ed. Richard Kearney (1985), p. 306

is quite clear that Newman's plans for the Catholic University from the start included the construction of an observatory and of a medical school. The observatory, which was inspired by similar institutions at the Vatican and Oxford University, never took shape, but a Dublin medical school located in Cecilia Street was purchased by the Catholic University and rapidly became a success. It was the base for much of the Catholic University's science teaching until the National University was founded in 1908.

Newman had advocated peaceful coexistence between theologians and scientists in his lectures, and displayed suspicion of the claims of Paley's natural theology which he believed left little room for 'real' theological discourse.[7] Had the Catholic church in reality been opposed to science or to scientific research, we would expect to find evidence of Catholic scientists who fell into disfavour with ecclesiastical authorities, or alternately who dropped lines of research as a result of ecclesiastical pressure. As we shall see later, the only case of this kind in Ireland in the period in question concerned not a scientist but a theologian.

Darwin and Catholicism in Ireland

Much of the published output of Irish Catholic scientific writers in the late nineteenth and early twentieth centuries can be labelled as apologetics, attacking either the idea that the church actively opposed science *per se* or the notion that science had effectively disproved many of the church's claims to knowledge. James R. Moore has characterised the reaction of Christians to Darwinism generally in terms of Leon Festinger's psychological theory of 'dissonance reduction'.[8] It is perhaps possible to detect three phases through which Irish Catholic writers progressed in adapting their faith to the new ideas.[9]

---

7   Discourses VII and VIII of Newman's *The Idea of a University, Defined and Illustrated* (1873), pp. 428–79. On the other hand John Wilson Foster ('Nature and Nation in the Nineteenth Century', in Foster and Chesney, *Nature in Ireland* (1997), p. 428–9) finds in Newman an 'impression of reluctance' when it comes to science.

8   Moore, *The Post-Darwinian Controversies: A Study of the Protestant Struggle to Come to Terms with Darwin in Great Britain and America, 1870–1900* (1979).

9   P.J. McLaughlin, 'A Century of Science in the *Irish Ecclesiastical Record*: Monsignor Molloy and Father Gill', *Irish Ecclesiastical Record*, 5th ser., 101 (Jan 1964), 34–40.

The first phase, of deliberately ignoring the problem as if it did not exist, can be characterised by Gerald Molloy (1834–1906), who held chairs in theology in the Royal College of St Patrick at Maynooth from 1857 and moved to UCD in 1873 as professor of natural philosophy and vice-rector. He became rector of UCD in 1883, and became well known as a lecturer and writer on science. His *Geology and Revelation*, based on articles originally written for the *Irish Ecclesiastical Record* and published in 1870 while he was still at Maynooth, attempts to reconcile the evidence of geology with the teachings of the Old Testament by either considering the 'Six Days' (Genesis 1:3–2:3) to have happened some time after the Creation or else allowing a less literal interpretation of the Hebrew word for 'day'.[10]

According to his colleague Walter McDonald, Molloy had intended to follow *Geology and Revelation* with a second volume dealing with the origin of Man. MacDonald suspected that 'having satisfied himself on many points, [Molloy] thought it more prudent to keep his conclusion to himself. He had no taste for martyrdom.'[11] Given MacDonald's own experience of 'martyrdom' described below, his suspicions are understandable. However, he was writing his own memoirs some forty years after Molloy's putative researches, and the attitude of the church to the origin of Man had changed considerably in those four decades.

The second phase, recognising that Darwinism existed but maintaining an outright hostility to it, was linked with the general phenomenon which Peter Bowler has called the 'Eclipse of Darwinism' around 1900. This was the widespread acceptance among scientists of the evolution of species, but without Darwin's mechanism of natural selection, driven instead by some general principle of 'progressive' development which many Christians found easier to reconcile with their concept of divine design. Indeed, Francis Darwin, Charles' son, chose his presidential address to the British Association meeting in Dublin in 1908 to proclaim his own belief in the inheritance of acquired characteristics rather than his father's theory. In Ireland, this period coincided with the resolution of the

---

10 Molloy, *Geology and Revelation: or the Ancient History of the Earth Considered in the Light of Geological Facts and Revealed Religion* (1870).
11 MacDonald, *Reminiscences of a Maynooth Professor* (2nd edn, 1967), pp. 73–4.

Irish University Question in 1908 and the consequent expansion of the University Colleges in Cork, Dublin and Galway as parts of the new, *de facto* Catholic, National University of Ireland.

University College Cork, from 1904 under the presidency of Bertram C.A. Windle, a prominent Anglo-Irish convert to Catholicism, became a centre of this kind of anti-Darwinian evolutionism. Windle, who had been professor of anatomy in Birmingham and had written on anthropology and archaeology, was a prolific propagandist on behalf of his adopted faith and against the evils of materialism and anti-religious science. Marcus Hartog, a leading supporter of vitalism since the 1890s, was already professor of zoology. Alfred O'Rahilly, who was to outdo Windle both in apologetics and in leaving his mark on the college as president, was appointed deputy professor of mathematical physics in 1915 (he became a full professor in 1917).

Windle wrote and edited several volumes addressing the relations between Christianity and science, most of which were published by the Catholic Truth Society. Perhaps his most substantial contribution was *The Church and Science* (1917), which addressed in its first two chapters the scope and limits of science and religion respectively, continued by surveying physics, astronomy, geology, early Man, vitalism and transformism (i.e. the transformation of species, with particular attention to Darwin and Mendel), and finished by considering Man and his origin, from first a scientific and then a religious point of view, without finding any irreconcilable difference between the two.

According to Windle, Darwin had successfully brought 'transformism' to the attention of the scientific mainstream, but the Darwinian mechanism – natural selection – could not account for all variation or transformation from one species to another; here Windle quotes approvingly the work of Gregor Mendel, who was after all an Augustinian monk. As far as Windle was concerned, the duty of the Catholic was to keep an open mind, particularly on the question of the origin of Man which science had yet to answer definitively. He outlined, with evident sympathy, St George Jackson Mivart's proposal that Adam had been created by God 'through the agencies of evolution acting on the lower animals', and noted that, although the church had discouraged the teaching of Mivart's theory, it had not censured or condemned it as heretical.[12]

The final phase, of not merely accepting the modern evolutionary synthesis once it had been established, but enthusiastically supporting it with theological arguments, was ushered in with the publication of H. de Dorlodot's *Le Darwinisme au Point de Vue de l'Orthodoxie Catholique*. De Dorlodot, who had held chairs in both dogmatic theology and stratigraphic geology in Belgian colleges, argued that there had been 'no special intervention of God for the formation of the world, beyond the creative action by which God drew the world from nothing, at the origin of time'.[13] This statement effectively drew a sharp line between theology and evolutionary biology. The former now had nothing to say about the precise evolutionary process which a practitioner of the latter might choose to advocate.

The importance of the evolution debate to Christian writers had decreased anyway after the First World War. The biggest perceived threat to Catholicism was no longer rationalism but socialism. The number of articles dealing with science published in Ireland's three main Catholic journals, the *Irish Ecclesiastical Record*, the *Irish Theological Quarterly*, and *Studies*, decreased after 1920 and the number of articles on politics both in Ireland and further afield increased. De Dordolot's book was reviewed in the *Irish Ecclesiastical Record* in 1922 by H.V. Gill, SJ, who had studied physics as a postgraduate in Cambridge under J.J. Thompson. This was the last article on evolution to appear in the *IER* for many years, although Gill continued writing for it on physics and historical topics.

In summary, although the Catholic church took about sixty years to accept that Darwinism was compatible with its teaching, this was little different from the scientific world as a whole. There is no evidence that at any point the Irish hierarchy used the threat of dangerous Darwinist ideas as an excuse for discouraging the teaching of science or for deterring research by Catholic scientists. If anything, the debate stimulated the publication of apologetics and of scientific treatises which supported the same side as the church in the

12 Windle, *The Church and Science* (1918).
13 Review by H.V. Gill, SJ, 'Catholics and Evolution Theories', *Irish Ecclesiastical Record* 19 (1922), 614–24.

ongoing debate about the mechanism of evolution. Once the scientific debate was settled and the modern synthesis of natural selection and genetics established, the church lost interest in the issue.

The McDonald case

There was one celebrated case of the Irish Catholic church constraining the publication by an Irish Catholic scholar of ideas which it found unacceptable. This was not, however, a case of geology, cosmology or Darwinist biology. The scholar in question was Walter McDonald, the prefect of the Dunboyne Establishment (effectively the graduate school of St Patrick's College, Maynooth), and his problems arose from his thoughts on the theological implications of dynamics, in particular the distinction between force and kinetic energy. McDonald believed that all causes of motion were directly activated by the will of God, and stated so in his 1898 book *On Motion*. His colleagues at Maynooth, and the Irish Catholic hierarchy in general, were concerned that McDonald's ideas endangered the doctrine of free will – if God directly activates every cause of motion, what role does that leave for human choice? – and had already requested him privately to desist from publicising his own ideas.

The bishops referred *On Motion* to Rome as soon as it was published. The book was swiftly placed on the *Index Librorum Prohibitorum*, and McDonald was ordered to withdraw all unsold copies. Although he complied with alacrity, five of the six books for which he subsequently sought a *nihil obstat* (permission to publish) during his lifetime were rejected. McDonald's unorthodox support of cattle-driving, and his belief that Catholics should participate fully in Trinity College, cannot have helped his case. He gained a posthumous revenge in an autobiography in which he cast severe doubts on the intellectual capacity of his opponents, at least in comparison with himself. One much later commentator in the *Irish Ecclesiastical Record* suggested that McDonald had 'sensed the importance of natural science but never learned its alphabet'.[14]

---

14 See W. McDonald, *Reminiscences of a Maynooth Professor* (1920). The detailed description of his *On Motion* (1898) and the controversy (pp. 111–66) is omitted from the second edition (1967). The final comment is from McLaughlin, 'A Century of Science', p. 253.

In conclusion, no particular theological deterrence to the pursuit of science emanated from the Catholic church, provided that scientists in Catholic institutions were orthodox or quiet about those issues of interest to theologians. McDonald, who was not a scientist but a theologian himself, might perhaps be balanced against the orthodox activists Molloy, Windle, O'Rahilly and Gill; but they can all be placed at the activist end of the scale of scientists' engagement with theology. Many other scientists in Maynooth, UCD, and the Cork and Galway colleges published nothing of much interest to theologians at all, and in this they were probably typical of most scientists of any nationality in their age.

15

The National University:
constructing Irish learning

The impact of the Irish University Question on Trinity College has been outlined in part I. In fact the various measures proposed and enacted had considerably greater effects on the other higher education institutions in Ireland. These were the Royal College of St Patrick at Maynooth, founded in 1795 by Act of Parliament as a Catholic seminary; the Queen's Colleges in Belfast, Cork and Galway, founded by Act of Parliament in 1845; and the Catholic University of Ireland, founded and funded by the Catholic church, which opened in 1854.[1]

Much of the nineteenth-century debate had revolved around the vexed question of the extent to which the state could be expected to sponsor 'denominational' rather than 'undenominational' or 'non-denominational' education. 'Denominational' in this context meant 'Catholic', since Trinity College had a distinct Ascendancy, Church of Ireland ethos and Queen's College Belfast inevitably

---

1 As before, I have used T.W. Moody, 'The Irish University Question of the Nineteenth Century', *History* 43 (1958), 90–109, and Fergal McGrath, 'The University Question' [1971], pp. 85–142. Also useful have been J.A. Murphy, *The College: A History of Queen's/University College Cork, 1845–1995* (1995); *Ollscoil na h-Eireann: The National University Handbook, 1908–1932* (1932); a number of essays in *Struggle with Fortune: A Miscellany for the Centenary of the Catholic University of Ireland, 1854–1954*, ed. M. Tierney (1954); and *A Page of Irish History: Story of University College, Dublin, 1883–1909*, compiled by Fathers of the Society of Jesus (1930).

developed a more dissenting and distinctively Ulster atmosphere. Most of the Catholic hierarchy wanted a church-controlled system of education for Catholic students. This, they argued, was the case in most European countries; the University of Louvain was often mentioned as a precedent. But successive British governments were hampered by the anti-Catholicism of their own supporters – if anything, more in England than in Ireland – and found it politically impossible to fund any project for Catholic education.

The Queen's Colleges had been founded in 1845 as constituents of the federal Queen's University of Ireland, a non-denominational institu-tion in which there were to be no religious tests and no teaching of theology. The Catholic hierarchy, after some initial hesitation and internal division, declared with the backing of Pope Pius IX that the colleges were dangerous to faith and morals and decided to set up their own university under the rectorship of John Henry Newman.[2] Without state financial support, the Catholic University was dependent on voluntary contributions, and without the legal power to grant degrees, it had difficulty in attracting students. In the meantime the hostility of the hierarchy to the three Queen's Colleges certainly slowed their development. On the one hand they were separated from the mainstream of Catholic nationalist Ireland, and on the other the Ascendancy perceived them as of lower status than Trinity College.

An institutional reform in 1879 brought some indirect state funding to the Catholic University. The Queen's University of Ireland was abolished and replaced with the Royal University, which was mainly an examining body. The new institution appointed twenty-six fellows who were to administer the examinations. Half of these fellowships were to be appointed from the staff of the Catholic University and Maynooth, and the others from the staff of the Queen's Colleges and Magee College, a Presbyterian theological college in Londonderry. Internal manoeuvres on the Catholic side resulted in Maynooth pulling out of the new arrangements and also in the Catholic University being renamed

---

2　Louis McRedmond, *Thrown Among Strangers: John Henry Newman in Ireland* (1990), and Fergal McGrath, *Newman's University: Idea and Reality* (1951).

University College Dublin, and passing under the control of the Jesuits rather than the bishops in 1883.[3]

This still left UCD in a parlous financial state, although it benefited greatly from the increase in state funding offered by the Royal University. The Queen's Colleges in Cork and Galway were still regarded with some suspicion by the hierarchy and this may have had some effect on their ability to attract students. It is clear that the great majority of students at Cork were there for the medical school.[4] The bishops, supported by Irish Nationalist MPs and by some Unionists, continued to lobby for the creation of a state-funded university which would be acceptable to them. There followed the two Royal Commissions of 1901–3 and 1905–6, which have been described in an earlier chapter, and the final settlement of the university issue by the creation of the National University of Ireland in 1908. This joined UCD with the renamed University Colleges in Cork and Galway, leaving Trinity College Dublin untouched and promoting Queen's College Belfast to separate status. Maynooth was permitted to affiliate to the NUI as well, and Magee College opted to join up with Trinity rather than QUB.

Although the bishops had consistently demanded formal control of any new Catholic university, in 1908 they settled for a situation where the church's control was more *de facto* than *de jure*. They might well have done so earlier, had such a solution been on offer from the government. In 1908 the general political atmosphere was perhaps more conducive to the compromise which was eventually imposed. Roy Johnston's accusation that the church carelessly ensured 'Protestant predominance in nineteenth century Irish science', by deterring Catholics from attending the Queen's Colleges, puts effect before cause. The failure to arrive at an acceptable format for Irish higher education was the result of a policy impasse between the Catholic church and the government, and it was after all the government, not the church, that had the power to legislate.

---

3   T.J. Morrissey, *Towards a National University: William Delany S.J. (1835–1924)* (1983), ch. 2, pp. 33–60.
4   Ronan O'Rahilly, *The Cork Medical School, 1849–1949* (1949), p. 24.

Science in the colleges

Even before the foundation of the National University, science was not restricted to the institutions of the Ascendancy and the state. Maynooth had a scientific tradition of its own, most notably in the work of Nicholas Callan (1799–1864), who produced one of the first electromagnetic induction coils in the 1830s. Although the study of science in the Catholic University of Ireland was not boosted by an observatory, as Newman had hoped, mathematics and physics were part of the general Arts curriculum, and the Cecilia Street Medical School became the base for the university's scientific research. John Tyndall, in his apologia for the Belfast Address, rather unconvincingly justified his speech as an attack on the Irish Catholic hierarchy in retaliation for their supposed refusal to allow the teaching of science at the Catholic University.[5] It has proved difficult to track down precisely what incident if any Tyndall was referring to. Certainly, the constraints on Cardinal Newman's beleaguered foundation seem to have been more financial than theological.

Under the Jesuits after 1883, UCD managed to attract and maintain a reasonable body of talent among its scientific staff. Most of them were attached to the Medical School, which was the most successful component of the Catholic University. Not all of the staff were Catholics; Thomas Preston, professor of experimental physics from 1891 to 1902, was part of George Francis Fitzgerald's Trinity College circle and the second recipient of the RDS's Boyle Medal, and his successor, J.A. McClelland, was an Ulster Presbyterian and a Cambridge graduate, who served as the Irish representative on the advisory council to the DSIR.[6] McClelland's strong research tradition on atmospheric ionisation was continued in UCD for many years after his death in 1920 by two brothers, J.J. and P.J. Nolan (1888–1952 and 1894–1984), who both became professors of physics in UCD. An early college history illustrates the constraints on science in UCD before the foundation of the Nationa

> The difficulties caused in the College for w,
> for teaching science can be hardly imagine

---

5   Tyndall, *Fragments of Science, vol. II*, pp. 212–14.
6   See DSIR reports.

McClelland's laboratory, which was used prior to the decease of Dr Molloy in 1906 [Molloy bequeathed most of his own equipment to the college]. Yet here was the only apparatus for preparing students even for the higher examinations in the University. We may relate a rather amusing incident in illustration of the severe handicap which happened in Professor McClelland's period. Two students from Cambridge who had studied with him under Thompson in the Cavendish Laboratory – which is celebrated as one of the finest in the world – were passing through Dublin on a tour, and calling at the College enquired for their distinguished fellow-student. Hearing that he was away at the moment, they asked the Dean of Studies, whom they happened to meet, if he would show them the present scene of McClelland's labours, expecting to see something worthy of the powers of research which they had known him to possess. The request was granted, and when they were shown into the class-room opposite the Sodality Chapel, with its few miscellaneous instruments huddled together in a couple of cupboards on the wall, they could scarcely conceal their astonishment. Quite unashamed, the Dean said, 'Gentlemen, I am glad to have given you some insight into the conditions under which our Professor of Physics has to work, so that when you return to your and his University you can say you have learned something new about the Irish University Question.' And no doubt they had.[7]

Queen's College Cork had been particularly well served by its second president, the chemist W.K. Sullivan, who had been head-hunted from the Catholic University in 1873. Until his death in 1890, Sullivan supervised an extension of the college's laboratory facilities with the considerable financial assistance of W.H. Crawford, a local brewing magnate.[8] The problem of the Catholic hierarchy's opposition to the Queen's Colleges remained, and the college declined under Sullivan's two uninspiring successors. But the arrival of Bertram Windle as president in 1904 and the resolution of the Irish University Question in 1908 enabled the college to attract larger numbers of students and indeed to diversify to the point where it was much more than a medical school with an Arts faculty attached.

---

e in Irish History, pp. 206–7.
ford also funded the construction of the college's observatory.

The Queen's College in Galway seems to have suffered from the drawbacks of its peripheral location and lack of laboratories. Its chair of natural philosophy had been held at various times by George Johnstone Stoney and Joseph Larmor, both of whom moved on to greater things fairly rapidly. Larmor's successor from 1885, Alexander Anderson, stayed in the post for almost fifty years, also serving as president of the Queen's College from 1899 and then of the University College from 1908. He was able at least to improve the laboratory facilities around the turn of the century through funding from Kodak – an unusual early example of direct industrial sponsorship of a university department.

On first sight, the prospects for science at the new National University when it was founded in 1908 should have been very hopeful. All three presidents of the constituent colleges and the university's chief administrator had scientific backgrounds – Windle in Cork, Anderson in Galway, and the president of UCD, Denis J. Coffey, a former professor of physiology and the dean of the Cecilia Street medical school. The new registrar of the National University, Sir Joseph McGrath, had been a lecturer in mathematical physics at UCD in the 1880s and early 1890s and had then been one of the secretaries of the Royal University.

However, the National University had been founded not to boost the teaching of science, but to create a higher education system in Ireland with which Irish Catholics were in sympathy. Although science fared no worse under the new dispensation than it had under the old, there was little expansion either. A few more posts were created – UCD attracted H.J. Seymour from the decaying Geological Survey – but the capital investment in laboratories and equipment which would have made a difference was not forthcoming from the state and does not seem to have been sought from private enterprise. Although UCD acquired the College of Science staff and buildings in 1926, this was only a rearrangement of the administration of existing resources.

Individual scientists with talent could and did flourish in this climate. A review of recent scientific research in UCD was published in the Jesuit journal *Studies* in 1921 and a bibliography of past and present members of staff and students of the National

University as a whole, including their scientific research, was published as part of the twenty-fifth anniversary volume in 1932.[9] The Nolan brothers have already been mentioned, and the fact that neither Arthur Conway nor Hugh Ryan was successful in parliamentary elections for the National University constituency[10] left them both free to produce numerous papers on electrodynamics and organic chemistry respectively. The Dairy School at UCC also seems to have produced an impressive number of publications.

Perhaps the most significant consequence of the foundation of the National University was the detachment of its constituent colleges from the academic job market which had previously included the whole of the British Isles. Most of the professors appointed to NUI colleges since its foundation have been NUI graduates, and the appointments system seems to have been drawn up in 1908 with the specific aim of favouring local applicants. At the time of the NUI's foundation, the Irish education system in general was being criticised by nationalists as being 'education for export', and perhaps the NUI was expected to reverse this by providing Irish jobs for Irish graduates. Whether or not this has been beneficial to the NUI's internal culture remains a fiercely contested question to this day.

---

9   James L. O'Donovan, 'Experimental Research in University College, Dublin', *Studies* 10 (1921), 109–22; and Ollsgoil na h-Eireann, *The National University Handbook, 1908–1932* (1932), pp. 215–70.

10 I quite sympathise. See Preface.

16

Science and Irish
nationalism

Another element in the scholarly consensus about the lack of science in Catholic Ireland has been the suggestion that Irish nationalists generally were too much taken up with the 'national struggle' or with the revival of Gaelic culture to be interested in scientific activities. There can be no doubt that the Irish cultural revival, and the Gaelic League in particular, did indeed appeal to the growing Catholic urban middle class in the decade or so after its foundation in 1893. According to Tom Garvin, 'the cultural atmosphere in which the new leaders [of the 1920s] had grown up was suffused with a nationalist and anti-modernist romanticism',[1] and John Wilson Foster supports Desmond Clarke's assertion that Irish nationalism and science were incompatible; the scientist, 'unlike the writer, artist or patriot, is rarely a mere nationalist but always has been and will be an internationalist'.[2]

But surely it is a false dichotomy to imply that good scientists cannot be 'mere' nationalists – Irish nationalists, from the context. Here it will be argued that this difference of outlook was not, and need not have been, a fundamental incompatibility, and that in fact

---

1   Garvin, 'Priests and Patriots: Irish Separatism and Fear of the Modern, 1890–1914', *Irish Historical Studies* 25 (1986), 67–81.
2   Clarke, 'An Outline of the History of Science in Ireland', *Studies* 62 (1973), 287–302, quoted in Foster, 'Nature and Nation in the Nineteenth Century', in Foster and Chesney, *Nature in Ireland*, p. 416.

the perceived hostility of Irish nationalists to science is to a certain extent a misinterpretation. It is important to understand the shifts in nationalist thinking over time. Around the end of the nineteenth century, a number of nationalists were very interested in scientific matters. The interaction of the cultural revival with science should not be simply characterised as the vampiric relationship which some writers seem to favour. The undeniable decrease in enthusiasm for state funding of science after independence is probably to be better understood as yet another facet of a general reluctance on the part of the new government to part with taxpayers' money for any purpose not perceived as essential.

Irish nationalists and science before 1920

The Irish Nationalist party under the leadership of Charles Stewart Parnell was as influenced by the Victorian industrialising milieu as any other group in the nineteenth-century United Kingdom. Parnell himself was an amateur mineralogist. His interest may have originated from his efforts to make his estate at Avondale, Co. Wicklow, profitable, but he went to the lengths of equipping his lover's house at Eltham with a small furnace so that he could carry out his own tests.[3] For vivid scientific imagery at the heart of Irish politics, it is difficult to better Tim Healy, during the December 1890 meeting in Committee Room 15 which sealed Parnell's fate, comparing the disgraced leader's personal magnetism to that produced by an iron bar in an electrical coil: 'This party was that electrical coil. There [indicating Parnell] stood the iron bar. The electricity is gone, and the magnetism with it, when our support has passed away.'[4]

In the years after Parnell's defeat and death, new voices began to be heard within Irish nationalism. One of the most influential of these was D.P. Moran, founder and editor of a weekly newspaper, *The Leader*, who became best known for the ideas put forward in his

---

3   As recounted by his widow Katherine O'Shea, *Charles Stewart Parnell: His Love Story and Political Life* (1914), vol. 1, p. 184. She also reveals that he was a fan of the popular astronomy works of Robert Ball (vol. 1, pp. 192–3); this enthusiasm was certainly not reciprocated by the astronomer. Edward Hull (*Reminiscences of a Strenuous Life*) also reports Parnell's interest in the minerals at Avondale.

4   Quoted by C. Cruise O'Brien, *Parnell and his Party, 1880–90* (1957), p. 318.

book *The Philosophy of Irish-Ireland*. He argued for the cultural revival to be linked with economic and industrial growth, and, perhaps with irony, praised the inflatable rubber tyre (invented by John Boyd Dunlop, a Belfast veterinary surgeon who was in fact from Ayrshire) as the only Irish innovation of any importance during the nineteenth century.[5] Moran provided a forum for a group of Catholic nationalist writers who supported the revival of the Irish language as part of a general revival of the strengths of the Irish nation. He and many other nationalists rejected all Irish Protestants as English colonisers, and the institutions of the Ascendancy – Trinity College and the Royal Dublin Society in particular – were regarded with suspicion and hostility as colonial bridgeheads.

However, this did not necessarily extend to an identification of the despised Ascendancy with scientific research. Sean Lysaght has examined the interaction between the Irish cultural revival and science at some length, concentrating particularly on the writings and speeches of W.B. Yeats – who, like Francis Darwin, addressed the 1908 meeting of the British Association in Dublin. Lysaght concludes that, although Yeats was 'of course' an opponent of science and the ideology of secular progress,

> [F]rom 1897 to 1908, we get an attenuation of Yeats's anti-scientific discourse. Against the background of Plunkett's reforming move-ment, Yeats's national ideal also extended to embrace the country's material prospects at this time. The case of Yeats is instructive here, since Yeats is so frequently upheld as the chief architect of the Revival, and the main exemplar of its otherworldly tone.[6]

From about 1908 the Gaelic revival became more political and also more sectarian. Yeats and a number of other Protestants withdrew or were expelled, and Moran's exclusive nationalism became more predominant. A new nationalist political party, Sinn Féin, was founded by Arthur Griffith, with the aim of an independent Ireland under the British crown which would emulate Germany by building up native Irish industry behind protective tariff walls. Although

---

5   Moran, *The Philosophy of Irish Ireland* (1905), p. 14.
6   Sean Lysaght, 'Robert Lloyd Praeger and the Culture of Science in Ireland: 1865–1953' (Ph.D. thesis, NUI. 1994), introduction.

building up native Irish industry behind protective tariff walls. Although Griffith's party collapsed a few years later (to be revived in the aftermath of the 1916 Rising), he succeeded in associating the slogan 'Sinn Féin' (We Ourselves) with a concept of separatist Irish self-sufficiency.

Although both Griffith and Moran in their different ways were committed to a largely romantic view of the Irish nation, both included the growth of Irish industry in their vision (and even Yeats was prepared to lean in this direction for a few years), and neither man could fairly be described as an enemy of the modern *per se*. The suggestion made by John Wilson Foster and others that the internationalism of science made it incompatible with nationalist political activity on the part of Irish separatists really does not hold water either. Even if the scientific endeavour is in any way truly 'international', this certainly does not mean that scientists are detached from their local political environment, and certainly the Ascendancy scientists whose political activities were described in part I were not so much internationalists as British nationalists.

One Irish political figure who had very firm views on the scientific enterprise was Arthur Lynch, a somewhat unusual member of the Irish Parliamentary Party in its closing years. He was born in Australia into an Irish emigrant family, fought for the Boers in the Boer War, and was first elected to parliament in the Galway city by-election of 1902 in which he defeated Sir Horace Plunkett. On arriving in England to take up his seat he was imprisoned for treason because of his military activity with the Boers. When he was released he became an MP again, this time for a Clare seat.

Lynch had also had some scientific training, and in his 1915 book, *Ireland: Vital Hour*, devotes an entire chapter to the Irish scientific genius. He echoes Havelock Ellis, and indeed the earlier polemics of Draper, in blaming nationalist Ireland's perceived lack of scientific achievement on the Catholic church, which 'in its relation to science, and in its treatment of men of scientific genius, has exhibited a tyranny only paralleled by its ignorance'.[7] In his later years, Lynch published books on 'the aletheian system of philosophy'

---

7   Lynch, *Ireland: Vital Hour* (1915), p. 328.

(1923) and on 'the case against Einstein' (1932). His successor as a parliamentary representative for Co. Clare was another politician with a scientific background, Eamon de Valera.

Certainly some among the rising generation of the 1900–20 period shared the 'nationalist and anti-modernist romanticism' that Garvin describes, but this was by no means true of all political activists, or even of all cultural activists. Another individual who has attracted the attention of both John Wilson Foster and Sean Lysaght is Michael Moloney, whose *Irish Ethno-Botany and the Evolution of Medicine in Ireland*, published in 1919, tries to reclaim the Irish roots of botany, herbal lore and medicine, and Lysaght gives other examples of the confluence of nationalism and science in the period.[8] Lynch's advocacy of science marks him as something of an eccentric, even in the odd grouping which was the Irish Parliamentary Party in its final years. But Moran's industrialising agenda became a part of the political platform of Arthur Griffith's refounded Sinn Féin after 1916, and as we shall see was adopted by Dáil Éireann before the independence of the Irish Free State.

Science in nationalist Ireland

The activities of a number of scientists who operated within the nationalist and Catholic education system have already been reviewed. However, the activities of the new Irish state's policymakers themselves in the years immediately before and after independence demonstrate that science *per se* was not completely ignored as a national priority. The history of the Dáil Commission of Inquiry into the Natural and Industrial Resources of Ireland shows that the revolutionary movement was indeed willing to look at scientific issues on the eve of taking power, and the writings of the new state's most influential educationalist clearly state both the priority he attached to scientific education and the idiosyncratic methods he advocated for it.

---

8   Moloney, *Irish Ethno-Botany and the Evolution of Medicine in Ireland* (1919); Lysaght, 'Contrasting Natures: The Issue of Names', in Foster and Chesney, *Nature in Ireland* (1997), p. 451; Foster, 'Nature and Nation in the Nineteenth Century', in Foster and Chesney, *Nature in Ireland* (1997), p. 421.

The Dáil Commission

Between 1919 and 1922, Dáil Éireann consisted of those Sinn Féin representatives who had been elected in the Westminster election of 1918 – after 1921, those elected in the Home Rule elections of that year. However, in the interpretation of Irish politics to which its members subscribed, it was in fact the parliamentary assembly for the entire Irish people. It was therefore responsible for carrying on the business of government as far as this was possible under revolutionary circumstances, despite many of the members of the Dáil and the executive being under arrest or on fundraising expeditions abroad.[9]

One element of the process of state-building was the creation of a Commission of Inquiry into the Resources and Industries of Ireland by the Dáil in June 1919. £5,000 was provided as funding for the commission, Darrell Figgis, a novelist and activist within Sinn Féin since 1914, was appointed its secretary, and sixty businessmen, scientists, industrialists and others of varying political backgrounds were invited to serve as members. Forty-nine accepted. But Dáil Éireann itself was declared an illegal assembly shortly after the commission was appointed, and the commission's attempts to hold public hearings were hampered to a greater or lesser degree by the efforts of the security forces to suppress it. This inevitably provided excellent propaganda for Sinn Féin, as government forces were publicly seen to be preventing peaceful discussions of economic development.

In spite of its difficult circumstances, the commission produced an interim report on the dairy industry in March 1920 and eventually produced seven more during the year starting from April 1921, on stock-breeding, coalfields, sea fisheries, peat, water power, dairying and industrial alcohol. The content of the reports is on the whole very good. Their subsequent impact on the government's policies after independence is less clear. Figgis moved on to write the Irish Free State's constitution during the first half of 1922, but by this time had parted from Sinn Féin and by 1923, as described earlier, was attacking the government for its lack of spending on science.

---

9   Much of what follows is drawn from Arthur Mitchell, *Revolutionary Government in Ireland: Dáil Éireann, 1919–22* (1995), pp. 80–5.

Timothy Corcoran, SJ

If there were some at least among the state-builders who were prepared to lobby for the interests of science, the same was true among educationalists. The most influential figure in the new state's education policies was Fr Timothy Corcoran, SJ, whom we have already encountered as a central figure in the National Academy of Ireland's brief existence. Corcoran was professor of the theory and practice of education in University College Dublin from 1908 to 1942, founded the journal *Studies* in 1912, was a member of all the relevant government commissions and committees, and wrote many articles on educational theory.[10]

Brian Titley characterises Corcoran's ideals as 'an extremely conservative Catholic view of education', viciously opposed to modern innovators such as Pestalozzi, Froebel and Montessori, very much in favour of repetitive teaching and examinations and the teaching methods of St John Bosco, St John Baptiste de la Salle, and of course the Christian Brothers. Corcoran's insistence that the teaching of any subject must be permeated with a Catholic ethos and if possible an Irish nationalist ethos is hardly surprising. Possibly his most damaging contribution to Irish education was to successfully persuade educators to teach as far as possible solely through the medium of Irish in order to realise the ideals of the Gaelic revival.

On the teaching of science, Corcoran's views were consistent with his central theme that memorisation and repetition, but not independent enquiry, were the keys to learning. So he believed that there had been too much emphasis under the previous regime on laboratory 'experiments' and not enough on training students to follow instructions, and that this had failed to deliver educational benefits:

> Manipulative skill is not the aim, or even necessary adjunct, of general education through science; the verification of certain processes in a laboratory is a useful consequence of such education, but by no means a foundation for it. The record of 'experiments' made in the student's notebook was made to replace and suppress

---

10 E. Brian Titley, *Church, State and the Control of Schooling in Ireland 1900–1944* (1983), pp. 94–100, reviews Corcoran's career and ideas. See also Anne Buttimer, 'Twilight and Dawn for Geography in Ireland', in Bowler and Whyte, *Science and Society in Ireland* (1997).

the use of a text-book. Training in the use of a text-book is the basis of all progressive education in science: and this reactionary plan of ousting the textbook, persisted in for many years, inflicted severe and permanent damage on Irish education and scientific progress . . .

What is the aim of a general or secondary education in Sciences? It is not discovery, it is not manipulative skill: it does not make the silly mistake of thinking that the function of science is realised in merely personal knowledge . . .

The function of the teacher in a general educative course of science is to demonstrate principles, indicate applications, guide students in attaining a grasp of what has been achieved, and direct them in the process of verification of knowledge previously presented and explained. It can be summed up as tradition and inspiration, with civic progress in view.[11]

Corcoran was not denying that scientific research could ever be carried out in the laboratory, quite the reverse; but he saw the university, not the secondary school, as the appropriate place for the teaching of the necessary research and technical skills. At some points his critique of the experimental process as taught in schools almost foreshadows Latour's seminal work on the social construction of scientific facts in the laboratory. However, Corcoran's commitment to a progressive and, though he would not have used the word, a positivist view of science, is firm.

Corcoran's policies on science teaching and compulsory Irish were implemented as far as possible in the 1920s, contrary to what has been stated by some recent writers. Michael Viney for instance says that in 1926 'rural science – the study of plants and flowers – was squeezed out of the national school syllabus to make more time for the study of Irish'.[12] The opposite is in fact the case. In 1926 rural science, which had previously been an optional subject, was made compulsory in schools where there was a teacher with the relevant qualification. It was in 1934 that the subject was made optional again with a view to expanding tuition in Irish, though in

11 Corcoran, 'The Place of the Sciences in General Education', *Studies* 12 (1923), 406–17.

12 Michael Viney, 'Woodcock for a Farthing: The Irish Experience of Nature', *Irish Review* 1 (1986), 62.

fact the shortage of suitably qualified teachers may have made little difference to the availability of rural science teaching.[13]

Alfred O'Rahilly

Bertram Windle's departure from the presidency of UCC in 1919 was virtually a forced ejection at the hands of his former protégé, the professor of mathematical physics Alfred O'Rahilly. Windle was a supporter of the old Nationalist Party, which disintegrated after 1918. His worst offence in the eyes of more advanced nationalists was to appeal to the British government, rather than to Dáil Éireann, for independent status for UCC. P.J. Merriman, the college's registrar, was appointed to succeed Windle as president. O'Rahilly took over as registrar, and enjoyed almost total control of the college as registrar (eventually from 1943 as president) for the next thirty-five years, retaining the chair of mathematical physics for the whole of that period.[14]

O'Rahilly was active in many fields. Apart from fulfilling his mission to make UCC a truly Catholic and Irish place of learning, he was one of Darrell Figgis' colleagues in writing the Irish Free State's constitution, served briefly as a TD for Cork Borough, campaigned for workers' education on Catholic principles, against communism, and against improper female attire in the college grounds. It is not very surprising that his output in the subject of his professorship was small. His biographer lists forty books and pamphlets by O'Rahilly, of which only four seem to be particularly related to science, and only one is at all connected to mathematical physics.

This was his 1938 volume on *Electromagnetic Theory*,[15] which challenged some of the fundamentals of Einstein's theory of special relativity in a characteristically robust manner. It is worth making the point that this critique, in particular its treatment of electric currents and of the relationship between electromagnetism and gravity, is still quoted by electrodynamic revisionists. Even though special relativity is largely accepted by today's physicist, there remain some areas of debate and O'Rahilly's ideas are part of that

---

13 E. Brian Titley, *Church, State and the Control of Schooling*, p. 136.
14 Murphy, *The College*, chs 7–11; J. Anthony Gaughan, *Alfred O'Rahilly, vol. I: Academic* (1986).
15 O'Rahilly, *Electromagnetic Theory: A Critical Examination of Fundamentals* (1965).

debate. Although one is inclined to assume that O'Rahilly's undoubted arrogance and utter confidence in his own abilities had got the better of him on this occasion, his scientific arguments rest on firm enough foundations – unlike, say, Arthur Lynch, whose *Case against Einstein* was mentioned above.[16]

Eamon de Valera and the Institute for Advanced Studies

A great deal of folklore still surrounds the academic achievements of Mr de Valera. Einstein is supposed to have said that only nine people, of whom de Valera was one, understood his theory of relativity. Indeed, more advanced admirers were known to claim that only two people, Einstein and de Valera, understood the theory and, in some parts of the country, there was considerable doubt about Einstein.[17]

The truth is of course that de Valera had a pass degree in mathematics from the Royal University of Ireland in 1904. He took a few courses at Trinity College in 1905–6, was professor of mathematics at Carysfort teacher training college from 1906 (having studied for a teaching diploma under Corcoran at UCD), and a part-time lecturer in Maynooth from 1912. He applied unsuccessfully for the chair later held by O'Rahilly at UCC in 1912, and had also applied for a post in UCG. Although his academic teaching career ended with the Easter Rising in 1916, he was chancellor of the National University of Ireland from 1921 until his death in 1975.[18] Throughout his life he maintained an intelligent though not a professional interest in quaternions, electrodynamics and tensor calculus, to which A.W. Conway of UCD and E.T. Whittaker of Trinity had introduced him.[19] Joe

16 For scepticism about O'Rahilly's 'refutation' of Einstein, see for instance Murphy, p. 274. Two of many recent articles sympathetic to O'Rahilly's case are H. Soodak and M.S. Tierstein, 'Dynamic Interpretation of Maxwell's Equations', *American Journal of Physics* 64 (1994), 907–13, and E.T. Kinzer and J. Fukai, 'Weber's Force and Maxwell's Equations', *Foundations of Physics Letters* 9 (1996), 457–61.

17 Donal McCartney, *The National University of Ireland and Eamon de Valera* (1983), p. 18.

18 McCartney; David Fitzpatrick, 'Eamon de Valera at Trinity College', *Hermathena* 133 (1982), 7–14.

19 Two anecdotes regarding de Valera's interest in science which I have been unable to research in depth must be relegated to this note:

Lee forthrightly summarises his contribution to the development of higher education in Ireland as follows:

> Chancellor of the National University of Ireland for fifty years, father of three professors in that university, the inadequacy of the resources his government devoted to higher education helped sabotage the pursuit of that intellectual excellence which would seem to be a prerequisite for fostering the cultivation of the mind that he was so fond of proclaiming as Ireland's mission in the modern world. Nevertheless, while neglecting all universities, he insisted that Trinity College should not be disproportionately neglected . . .[20]

De Valera's most significant contribution to science as a politician was undoubtedly the foundation of the Dublin Institute for Advanced Studies in 1940 with its Schools of Celtic Studies and Theoretical Physics, to which Dunsink Observatory was added as a School of Cosmic Physics in 1947. This falls somewhat outside our time-span, but it is worth reviewing the reasons why theoretical physics was chosen as a topic of study at the new institute. The most immediately apparent explanation is that the institute's specialisations were de Valera's own particular enthusiasms. He removed his minister for education from office and took on that portfolio himself in 1939–40 so that he could manage the parliamentary passage of the bill creating the institute.[21] Another factor must have been de Valera's desire to find a post for the physicist Erwin Schrödinger, whom he had persuaded to move to Dublin in 1939 with his household.

---

19 (*cont.*) i) Professor Hermann Brück described to me (in an interview in 1992) an apparatus of wires and wooden balls which de Valera used to interpret some particular aspect of mathematics which he was interested in – this must have been in the early 1950s when his vision was becoming increasingly impaired. It would be interesting to establish what the apparatus was used for and indeed if it survives.

ii) Dr Sean Faughnan informs me that scribbled in the margins of the drafts of the 1937 Irish Constitution, which was handwritten by de Valera, are a number of mathematical expressions which from his description seem likely to be some form of tensor calculus. It would be interesting to know what else was on the statesman's mind as he rewrote his country's basic law.

20 Lee, *Ireland 1912–85* (1989), p. 332.

21 This interpretation is also supported by Lee, p. 238.

However, he could hardly offer either personal whim or a desire to cater for a foreign scientist, no matter how distinguished, to the Dáil as a justification for the institute. Instead he appealed to the historic Irish mathematical tradition, and also pointed out the comparatively inexpensive nature of theoretical physics as 'a branch of science in which you want no elaborate equipment, in which all you want is an adequate library, the brains and the men, and just paper'.[22] Ireland could not afford massive investment in laboratory equipment, but it could provide books and desks.

The Irish State and Science

In earlier chapters we saw how the Ascendancy institutions and the publicly-funded institutions of science declined under the Irish Free State's government, and here we have seen that the National University did not particularly gain from the country's new status either. However, the failure of the new state, under governments of whatever party, to innovate in education and science was matched by its reluctance to innovate on other issues such as economic policy, employment, health, or indeed Partition. This was caused by the tight control on public spending exercised by the Department of Finance, and by that department's enthusiasm for cutting public spending whenever possible.

The foundation of the Institute for Advanced Studies was one exception to this trend. Two other investments of public funds for scientific or engineering purposes which shall be noted here were the Shannon hydroelectric power scheme, initiated over the head of the Department of Finance in 1924, and the trains which ran on a local Dublin line between 1932 and 1949 driven by an electric battery designed by the Irish inventor James Drumm. The state was as little inclined to spend money on science as on anything else, but a determined or cunning lobbyist could still break through the barriers of fiscal conservatism.

Nationalist science

In summary, then, the case against either Irish nationalism or Roman Catholicism as despoilers of science remains to be proven. In

---

22   *Dáil Debates* 76 (6 July 1939), 1969–70.

particular, the Catholic church's role in the promotion of science in Ireland was probably as positive as financial and political constraints allowed. Debate of the latest scientific ideas even flourished in ecclesiastically sponsored journals such as the *Irish Ecclesiastical Record*, the *Irish Theological Quarterly*, and from 1912 Corcoran's *Studies*. Apart from the case of McDonald, who was not a scientist, there is no evidence of the church restraining or attempting to restrain scientific research for theological reasons.

The case against nationalism is a little more complicated. It is easy to characterise the entire nationalist movement as anti-modern and anti-materialist, but in fact several influential figures within it did concern themselves with the agenda of development and of science, and before the formal granting of independence they were relatively successful in promoting it within their movement. It was after 1921 that opportunities were not grasped by policy makers, and it is perhaps the romanticised ideal of the Irish nation which was established as the post-independence norm that has enabled subsequent historians to forget that some nationalists were modernisers. De Valera, in the election year of 1943, set out what he insisted was the true vision of Irish nationhood. Even he must have already known that it would remain a dream of the past rather than the future:

> That Ireland which we dreamed of would be the home of a people who valued material wealth only as the basis for right living, of a people who were satisfied with frugal comfort and devoted their leisure to the things of the spirit – a land whose country side would be bright with cosy homesteads, whose fields and villages would be joyous with sounds of industry, with the romping of sturdy children, the contests of athletic youths and the laughter of comely maidens, whose firesides would be forums for the wisdom of serene old age. It would, in a word, be the home of a people living the life that God desires that man should live.[23]

---

23 Tim Pat Coogan, *De Valera: Long Fellow, Long Shadow* (1993), p. 628.

17. Conclusion

The preceding chapters have examined in some detail the surviving evidence of the social and political context of science in Ireland. We have seen how three different strands of activity can be identified within the history of Irish science. The time period with which we are mainly concerned, between 1890 and 1930, is unique in that all three of these scientific strands of activity were active to roughly the same degree. Ascendancy science, after a century or so of activity and a decade or two of gradual relaxation, was about to hit the cataclysm of the treaty; administrative science saw some decades of growth crowned with the foundation of the Department of Agriculture and Technical Instruction, only to find itself edged out of the state-building programme of the new government; and nationalist science received its first substantial recognition by the state in the foundation of the National University.

Despite the claims that are sometimes made for the objectivity and internationalism of science, we have seen that for each of these three groups science was inseparable from its social and political context. For the Ascendancy it was a proof of the intellectual superiority of their tradition and for some a bulwark against Roman Catholic superstition. Administration scientists brandished Irish separatist or British imperialist agendas at each other as they tussled over unique specimens. Some nationalists saw applied science as the key to Ireland's future industrial regeneration. Others preferred instead to

183

sponsor the more cerebral (and less expensive) realms of mathematical physics. In the first issue of the *Irish Review* in 1986, Dorinda Outram asked what was specifically Irish about the scientific enterprise on the island; we have come some way to answering her question.

This concluding section will address three issues which are germane to all three of the strands which have been described. First, is there evidence that science declined in the Irish Free State during the first decades of independence, and if so, does that evidence support any of the explanations put forward here or by other writers? Second, can we shed any light upon the apparent neglect of science by Irish historians and other cultural scholars in more recent decades? Third, to what extent has the concept of 'colonial science' been useful here, and to what extent therefore can Ireland be described as a colony of Britain at least in terms of the history of science?

Science in a Free State

The perceived failure of science in the first few decades of the Irish state's existence is a recurrent theme of current writings on the history of Irish science. The difficulties of the political transition for each of the strands have been described: the marginalisation of Trinity College and the Royal Dublin Society; the public spending squeeze which hit the National Museum and the Geological Survey; the promises of nationalism for science which were never kept. Two factors which applied to all three traditions more or less equally have been mentioned in all of these cases, but deserve to be repeated here.

The first is the simple effect of almost ten years of war, the World War from 1914 to 1918, the War of Independence from 1919 to 1921, and the civil war of 1922–3. While war stimulated John Joly to ever more ingenious inventions of potential military application, in general its effect on science and society in Ireland was very disruptive indeed. The two young geologists, Valentine and Kennedy, and the biologist Selbie, were actually killed, but others left during and after the treaty settlement. The financial base of the Irish economy was depleted by the exodus of Protestants in the early 1920s; the intellectual base for the growth of Irish science was hit both by war casualties and by emigration; and the industrial base, too, was affected by the relocation of some firms to England (such as the Grubb Parsons engineering works, which had created so many telescopes) and by the reluctance of others to invest in the first place in a new state surrounded by tariff barriers. None of the three strands of Irish science was unaffected by these events. If they have not been emphasised in previous chapters, it is simply because the story is almost the same in each case.

The second is the reluctance of the new state to spend money on anything. As noted in the previous paragraph, the Irish tax base had been seriously depleted by 1923. Ireland had been a net contributor to the UK treasury in the 1890s, but by the 1920s the situation had reversed, so much so that the arrangements for the financial autonomy of Northern Ireland made in the 1920 Government of Ireland Act were found to be unworkable within a few months of their implementation, and the Irish Free State could reasonably claim a partial victory in the 1925 Boundary Commission settlement because the new state was released from its share of the United Kingdom's national debt as part of the deal. The fiscal caution of the first decades of Irish independence is legendary, and led in the end to the ejection of Cosgrave's government in 1932 after it had cut the old-age pension. This constraint on the public purse had its effect on science, of course, but science suffered only as much as any other branch of expenditure, and certainly less than primary school education which was forced down the blind alley of universal compulsory teaching in the Irish language, or public housing which remained awful until the mid-century. We should not miss the wood of the conservatism of the Department of Finance in the trees of the underfunding of Trinity College, the National University, and the Science and Art Institutions.

The decline of scientific activity after independence does have one important implication for historicist theories applying the colonial model to Irish history. The third, 'nationalist', phase of science is presented by both Krishna and Basalla as the acme of cultural achievement, providing a secure platform for future scientific development and advance. In Ireland this was not immediately the case. Indeed, the events leading up to independence affected the nationalist tradition just as adversely as the other two, and apart from the transfer of the Royal College of Science to UCD in 1926 there was little advance for the relative position of nationalist science until the 1940s. There is nothing inevitable about the onward march of science. External factors, political, military, or economic, can affect the context in which science is practised so as to make it virtually impossible.

Science in (or out of) Irish culture and history

The absence of science until very recently from general discussions of Irish culture and history is undeniable. Perhaps it is because, as Gordon Herries Davies put it, 'our historians have felt more comfortable in discussions of banking, battles and bishops than in dealing with problems concerning basalt, binomials and brachiopods'. Perhaps it is because Irish culture was constructed – particularly by W.B. Yeats – as an anti-modern, Romantic construct with no room for the empirical rationalism which science is so often described as possessing. Perhaps it was because many actors identified science in Ireland with the Ascendancy culture which had nurtured it for so long and it was therefore alienated from the new state, as we saw in the case of the Royal Dublin Society's transition.

But there may be a less Irish reason for this characteristic of Irish studies. It is in general across the world rather unusual for scientists to take much part in the mainstream discourse of the humanities, and this is because the sciences and the humanities are in fact different fields of discourse in the first place. True, the history of science is a weak and late arrival on Irish academic soil, but so are a number of other disciplines. It is reasonable to raise the question of why the history of science has not been encouraged to grow on the island, but to understand why this has been so, we need also to compare its lack of growth with that of other subjects taught (or not taught) in Ireland, or with the history of science as it is taught elsewhere. There is generally believed to be a distinction between science and the arts, and while we may regret this it should not surprise us.

Of course, this distinction is not absolute. On several occasions, Irish science has benefited from an explicitly cultural agenda. We have seen how the Royal Irish Academy was preserved from the threat of Timothy Corcoran's National Academy by invoking its historical contributions to the study of Irish culture – which were at their peak of activity just as Irish astronomy and mathematics were flourishing under the RIA's auspices. The 'constructive Unionist' agenda implemented by Sir Horace Plunkett gained the support of W.B. Yeats and the Gaelic revival for a few crucial years as it got under way. And de Valera's Dublin Institute for Advanced Studies combines departments of Cosmic and Theoretical Physics with a department of Celtic Studies.

Happily in recent years the Irish studies community has become more receptive to the idea of Irish science as a part of Irish culture. Issue 20 of the *Irish Review*, published in the spring of 1997, was the third in succession to include an article contributing to the debate on the location of science in Irish cultural space. The interest of the Cultural Traditions Group in Northern Ireland in the history of science has already been (gratefully) acknowledged. There has been some public interest in summer schools commemorating John Tyndall and George Gabriel Stokes, and the latest step in the restoration of the great telescope of Birr Castle features regularly in newspaper reports. The prospects for a more positive engagement of science and culture in Ireland at the turn of the millennium seem very favourable.

Colonial science and Irish science

The final issue to be considered here is, to what extent the concept of 'colonial science' has been useful in this survey of the social context of science in Ireland between 1890 and 1930, and to what extent therefore Ireland can be described as a colony of Britain, at least in terms of the history of science. This has wider implications. As we saw in the introduction, Liam Kennedy has argued vigorously that as an economic historian, there is no way that he can see Ireland as a post-colonial state; on the other hand, Declan Kiberd has argued equally vigorously that Irish literature can be understood only in terms of the post-colonial nation-building project.

This study has adapted the models proposed by Krishna and Basalla for the spread of science into the particular colonial contexts of India and North America. Roy MacLeod, looking at the slightly different cases of Canada, New Zealand and Australia, found a five-stage taxonomy more appropriate to these countries. The first lesson to be learnt here is that every situation is unique, and that a model which helps us understand the context of science in any one country or region will certainly require modification and may be completely useless if we try to apply it to another.

Why should colonial models be so relevant to the Irish experience? Is not Ireland after all a European nation, and a region of the British Isles, and should we not seek European national comparisons or British regional models of scientific development? To take the European case, Ireland is unusual among the European nations which gained independence in the first quarter of this century in two respects. It had had no autonomous organs of government since 1800 (unlike Finland or Norway), and its history before independence had been one of connection with only one powerful neighbour (unlike the new states in Central and Eastern Europe, which were crushed between several great powers, or Norway which had been a more equal partner in its various Scandinavian linkages). In that respect Ireland's historical situation can indeed be usefully compared with that of the Indian sub-continent or with the Dutch East Indies, or even with those South American countries which though politically independent fell under the economic and cultural hegemony of one of the European powers.

There are two important reasons for not treating Irish science simply as a special case of British regional science. The first is that, unlike any other British region, Ireland became detached from Westminster government in the early 1920s as the Irish Free State gradually evolved to full independence and Northern Ireland too acquired an autonomous regional administration. The second is that there existed a fundamental cultural divide between Protestant Ascendancy and Catholic separatism, which was more deeply entrenched – and in somewhat altered form remains so in Northern Ireland – than in any other region of the United Kingdom. This is one aspect of the problem where a comparative European perspective is helpful. The divisions among Finnish scientists along linguistic lines occurred at almost the same time as these issues arose in Ireland, and there are similar divisions elsewhere in Europe.

This last point is also the biggest difference between the colonial situations referred to above and the Irish situation. For both Basalla and Krishna, the first category or phase of scientists to become active were outsiders from Britain, coming to America or India to study its plants, animals, rocks, people, and then reporting back home. In Ireland this elite was a native one with its own distinctly Irish scientific tradition, which certainly went back to the eighteenth century and could arguably be pushed back considerably earlier to Robert Boyle, Hans Sloane, or even Archbishop Ussher. Although the Ascendancy is often described as 'colonial', this is not a very helpful description in terms of understanding their role in Irish society. It does of course accurately describe the many members of the Ascendancy – and of other Irish backgrounds – who participated in the British imperial enterprise.

Irish Catholic scientists too began at an advantage with respect to their 'native' counterparts in India or the Dutch East Indies. They were physically indistinguishable from their Protestant fellow-countrymen; they shared the same vernacular; they had some access to higher education, unsatisfactory though it was until 1908, and even before then a few, such as the first two presidents of Queen's College Cork, Sir Robert Kane and William K. Sullivan, were able to push past the gatekeepers and join the Irish scientific elite. That these were the exceptions rather than the rule indicates the exclusive nature of Ascendancy science; but that they existed at all demonstrates the difference between Ireland and other situations.

The closest and fairest point of comparison with colonial science is the group which I have labelled 'administration scientists'. These men (and occasionally women) were mostly English by birth, state employees (by definition) and trained to see Ireland as an object of study. Yet we have seen how, despite their backgrounds, both Grenville A.J. Cole and R.F. Scharff 'went native' at different times and used the arguments of Irish nationalism to further their own scientific and institutional agendas. More than either of the other two groups, the experience of the administration scientists demonstrates that the point of reference for their scientific work was moving between South Kensington and Leinster House; the 'moving metropolis' which Roy McLeod has identified in the examples he has studied.

In summary, then, the story of the history of science in Ireland is unique, just as the history of science in any nation or region is unique. Irish science was never a unified, homogeneous phenomenon. Major Denis Pack-Beresford, who divided his energies between collecting insects and writing Unionist pamphlets from his country seat, inhabited a different world from Hugh Ryan, fiercely nationalist professor of chemistry at University College Dublin. Irish culture was divided, at times bitterly, between the cultures of more-or-less Catholic nationalism, Ulster Protestantism, and smallest of all the Anglo-Irish Ascendancy, the class from which many Irish scientists came. All claimed fervently to be Irish and to have their country's best interests at heart; but Pack-Beresford's Ireland was not Ryan's. Although the other situations which have been of most use as points of comparison with Ireland are on the whole those of former British colonies across the world, Ireland is nonetheless a European nation rather than a British colony. The case for treating Irish science as an example of colonial science can be made, but it is not proven.

Appendix 1

Background, institutional affiliation and ranking of candidates for the Royal Dublin Society Committee for Science and its Industrial Applications, 1915–25

Name	Background	Institution	1915	1916	1917	1918	1919	1920	1921	1922	1923	1924	1925	Average Rank
JP Drew	?	UCD						~~16~~	2					9
H Pringle	?	TCD								13	14	7	11	11.25
GAJ Cole	E	S+A	2	3	2	2	11	6	7	2	10			5
GT Morgan	E	RCSc	7											7
GH Pethybridge	E	DATI	12	9	5		5	6	8	11	8			8
JA Scott	E	Med		15	6	12	7	10	12	6	4	5	6	8.3
JW Parkes	E	Ind						~~20~~	~~19~~	~~17~~	~~17~~	10	13	16
KC Bailey	E	TCD										~~16~~	~~16~~	16
HC Plummer	E	TCD	~~20~~	~~19~~	~~18~~	~~16~~	~~16~~							17.8
AR Nichols	E	S+A			~~19~~	~~20~~								19.5
D Clark	E	TCD								~~20~~				20
J Wilson	Sc	RCSc	3											3
W Brown	Sc	RCSc	4	4	4	3	8	5	13	11	13	9	8	7.5
F Moore	Pr	S+A									1	1	1	1
AF Dixon	Pr	TCD	1	2										1.5
JW Moore	Pr	Med	5	1	1	1	1	1	1	1	3			1.7
EA Werner	Pr	TCD	9	6	8	6	3	4	8	4	6	3	3	5.5
EC Bigger	Pr	Med										2	9	5.5
G Fletcher	Pr	DATI		13	3	4	15	3	6					7.3
AGG Leonard	Pr	RCSc			11	9	4	13	15	9	8	6	4	8.8
JHJ Poole	Pr	TCD										13	5	9
NG Ball	Pr	TCD							15	5	7	9		9
R L Praeger	Pr	S+A	11	10	9	10	12	14	10	10	7	8	7	9.8
W Kaye-Parry	Pr	Ind	6	5	14	8	13	12	14					10.3
WH Thompson	Pr	TCD	15	8	15									12.7
JB Gatenby	Pr	TCD								15	12			13.5
LB Smyth	Pr	TCD							~~17~~			14	10	13.7
WJ Buchanan	Pr	Med									15			15
JL Synge	Pr	TCD											15	15
JT Wigham	Pr	TCD	~~18~~	~~16~~	~~19~~	15	14							16.4
C Green	Pr	DATI		~~20~~	~~17~~									18.5
WB Wright	Pr	S+A		~~20~~	~~17~~									18.5
HJ Seymour	RC	UCD	10	11	7	7	5							8
A Henry	RC	RCSc	~~17~~	~~17~~	~~16~~	14	10	2	4	5	2	4	2	8.5
PF Purcell	RC	RCSc	~~16~~	7	10	13	2	9	3	3	11	11		8.5
FEW Hackett	RC	RCSc	13	12	13	11	9	7						10.8
JB Butler	RC	UCD	8	~~18~~	~~20~~	~~17~~	~~18~~	11	9	14	5	12	14	13.3
JJ Nolan	RC	UCD						~~19~~	~~16~~	12	~~16~~	15	12	15
H Ryan	RC	UCD	14	14	12	~~18~~	~~19~~	~~17~~						15.7
H Kennedy	RC	UCD						~~18~~		~~16~~				17
PA Murphy	RC	DATI									~~18~~	~~17~~	~~17~~	17.3
JM O'Connor	RC	UCD						~~18~~	~~18~~					18
J Reilly	RC	Chem										~~18~~		18
TJ Nolan	RC	Chem										~~19~~		19
EJ Sheehy	RC	RCSc										~~19~~		19
T Hallissy	RC	DATI						~~20~~		~~19~~	~~18~~	~~19~~		19
LS Gogan	RC	S+A										~~20~~		20
HA Lafferty	RC	DATI									~~20~~			20

Backgrounds. Sc: Scots, E: English, Pr: Irish Protestant (inc. Quaker), RC: Irish Catholic, ?: unknown.

Institutions. TCD: Trinity College Dublin, UCD: University College Dublin, Ind: industry (Parkes: chemist, Parry: engineer), Med: doctor or surgeon, RCSc: Royal College of Science, S+A: other Science and Art (Geological Survey, Botanic Gardens, National Museum, National Library), DATI: other Department of Agriculture, Chem: State Chemist.

Rankings. Unsuccessful candidatures (ranked lower than 15th) are ~~struck-through~~.

Source. RDS records.

Appendix 2

Members of the RIA Council for Science elected in the years 1890–1939

Trinity College Staff:
K.C. Bailey (1927, 1930–2, 1936–7)
(Sir) Robert S. Ball (1890–2)
J.W. Bigger (1931, 1935–6)
C.R. Browne (1902–4)
G.L. Cathcart (1890–1901)
J.R. Cotter (1935)
D.J. Cunningham (1890–3, 1900, 1902)
R.W. Ditchburn (1933–8)
A.F. Dixon (1906, 1914–15, 1918–20, 1923–6, 1929, 1933)
W.R. Fearon (1924–5, 1928, 1934)
M.W.J. Fry (1914–6, 1919–21, 1925–7)
J.B. Gatenby (1927)
Samuel Haughton (Pr. 1886–90; Co. 1891–6)
Charles J. Joly (1898–1905)
John Joly (1903, 1915)
A.J. McConnell (1932–4, 1938)
T.G. Moorhead (1936–7)
A.C. O'Sullivan (1921–3)
Henry C. Plummer (1921)
J.H.J. Poole (1935–9)
H. Pringle (1925)
F. Purser (1906–9)
A.A. Rambaut (1895–7)
C.H. Rowe (1931–3)
R. Russell (1921)
Louis B. Smyth (1931–2, 1938–9)
W.J. Sollas (1892–7)
J.L. Synge (1927–30)
F.A. Tarleton (Co. 1890–1917, inc. Pr. 1906–10)
W.E. Thrift (1919, 1922–4)
W.R. Westropp Roberts (1902–1929)
Edmund T. Whittaker (1908–1911)
B. Wilkinson (1890–1901)
E.P. Wright (1890–9)
Sydney Young (Co. 1907–10, 1912–14, 1917–18; Pr. 1921–5; Co. 1926–7)

Connected to Trinity College:
Fourth Earl of Rosse (Co. 1894–1901, inc. Pr. 1896–1900)
C.R.C. Tichborne (1890–1)

Staff of the Royal College of Science:
W.E. Adeney (1904, 1906–8, 1910–11, 1913–14, 1918–20, 1923)
G.H. Carpenter (1907–9, 1911–13, 1917–18, 1921–2)
Grenville A.J. Cole (also Director of Geol Survey from 1905) (1897–1905, 1907–10, 1912–13, 1917–18)
Felix E.W. Hackett (1920–2)
Augustine Henry (1920–1, 1924–6)
Thomas Johnson (1902–4)
J.P. O'Reilly (1890–8, 1901–3)

Other Science and Art Institution employees:
Valentine Ball (1891–4)
Robert F. Scharff (1900–1, 1903–15, 1908–11, 1913–5, 1919–20, 1923–6, 1929–31)
T. Hallissy (1925–7)
G.H. Kinahan (1890–1900)
(Sir) F.W. Moore (1898–1902, 1905–8, 1910–12, 1915–17, 1920–4, 1926–30, 1933–8)
Robert Lloyd Praeger (Co. 1903–39, inc. Pr. 1931–3)
A.W. Stelfox (1928–9)

Other DATI employees:
G. Fletcher (1913–14, 1918–19, 1928)
G.H. Pethybridge (1916–17, 1921–3)

University College Dublin:
J. Algar (1932–4, 1939)
J. Bayley Butler (1923, 1936, 1937)
J. Casey (1890)
Arthur W. Conway (Co. 1912, 1916, 1919, 1922, 1927, 1928–31, 1934–5; Pr. 1937–9)
E.J. Conway (1939)
J. Doyle (1931–3, 1938–9)
Felix E.W. Hackett (1926, 1929–39)
Augustine Henry (1926)
J.A. McClelland (1906–20)
E.J. McWeeney (1904)
J.J. Nolan (1922–39)
P.J. Nolan (1932–4, 1939)
T.J. Nolan (1928–31, 1934–5, 1939)
Hugh Ryan (1913–15, 1918–19, 1922, 1925, 1930)
Henry J. Seymour (1932)
George Sigerson (1890)
F.T. Trouton (1899–1902)

President of University College Cork:
Bertram C.A. Windle (1916–17)

Professor at University College Galway:
T. Dillon (1937–8)

Professors at Queen's College Belfast/Queen's University of Belfast:
Gregg Wilson (1904–6)
J.K. Charlesworth (1924)
W.B. Morton (1930–2, 1935–9)

Scientists with an independent income:
R.M. Barrington (1905–7)
John Ellard Gore (1905–7, 1909–10)
Denis R. Pack-Beresford (1920)
Greenwood Pim (1896–1902)
Sir John Ross of Bladensburgh (1909–12, 1915–17)
R.J. Ussher (1911–2)
William E. Wilson (1903–6)

Others:
C.F. D'Arcy, Archbishop of Armagh (1924)
A.E. Mettam, Principal of the Royal Veterinary College (1914–16)

Years of election, not terms of office, are given in all cases. Years of election as president (Pr.) are given as well as years elected to council (Co.) for Haughton, Tarleton, Young, Rosse, Praeger and Conway.

Source: RIA Proceedings.

Appendix 3

Proposals of the 1920 Royal Commission on Trinity College

Subject	1920 annual expenditure	Proposed expenditure	Proposed increase	Proposed capital grant
Mathematics	£1,170	£1,170	± £0	
Physics	£2,400	£3,850	+ £1,450	
Dunsink Observatory	£1,300	£2,200	+ £900	
Chemistry	£2,272	£4,500	+ £2,228	£31,500
Botany	£1,950	£4,000	+ £2,050	£2,000
Geology	£1,050	£2,450	+ £1,400	
Zoology	£900	£3,050	+ £2,150	£7,000
Classics	£1,600	£2,000	+ £400	
Oriental Languages	£880	£1,450	+ £570	
Romance Languages	£770	£1,950	+ £1,180	
Germanic Languages	£435	£1,350	+ £915	
Slavonic Languages	£60	£350	+ £290	
Celtic Languages	£240	£1,550	+ £1,310	
English Literature	£425	£900	+ £475	
Philosophy	£130	£1,300	+ £1,170	
Economics	£350	£2,750	+ £2,400	
History	£1,400	£3,250	+ £1,850	
Divinity	£600	£700	+ £100	
Law	£1,150	£1,600	+ £450	
Medicine: Surgery	£430	£950	+ £520	
Medicine: Anatomy	£2,078	£3,062	+ £984	
Medicine: Pathology/ Bacteriology	£2,050	£4,400	+ £2,350	£16,00010
Medicine: Physiology	£1,606	£3,384	+ £1,778	£19,500
Engineering	£2,130	£6,400	+ £4,270	£25,000
Agriculture	£0	£2,000	+ £2,000	
Assistant Registrar	£300	£400	+ £100	
Library	£3,932	£5,520	+ £1,588	£12,400
Women Students	£475	£900	+ £425	
Research Studentships	£0	£2,400	+ £2,400	
General Expenditure	£0	£12,000	+ £12,000	
Total	£32,083	£81,786	+ £49,703	£113,400

Source: Report of the Royal Commission on the University of Dublin, 1920 (Cmd 1078).

Bibliography

A. Books and articles:

Peter Alter, *The Reluctant Patron: Science and the State in Britain, 1850–1920*, tr. by Angela Davies (Oxford: Berg, 1987)

R.A. Anderson, *With Horace Plunkett in Ireland* (London: Macmillan, 1935)

C.S. Andrews, *Dublin Made Me: An Autobiography* (Dublin & Cork; Mercier Press, 1979)

David Attis, 'The Social Context of W. R. Hamilton's Prediction of Conical Refraction', in Bowler and Whyte, *Science and Society in Ireland* (1997), pp. 19–36

Kenneth C. Bailey, *A History of Trinity College, Dublin, 1892–1945* (Dublin: The University Press/Hodges Figgis, 1947)

Robert S. Ball, *In Starry Realms* (London: Isaac Pitman, 1892)

— *Great Astronomers* (London: Isbister, 1895)

— *Reminiscences and Letters of Sir Robert Ball*, ed. W. Valentine Ball (London: Cassell, 1915)

Jonathan Bardon, *A History of Ulster* (Belfast: Blackstaff, 1993)

Jacques Barzun, *Race: A Study in Modern Superstition* (London: Methuen, 1938)

G. Basalla, 'The Spread of Western Science', *Science* 156 (1967), 611–22

J.C. Beckett, *The Anglo-Irish Tradition* (Belfast: Blackstaff, 1976)

— *The Making of Modern Ireland 1603–1923* (London: Faber & Faber, 1966)

D.E. Beesley, 'Isaac Ward and S Andromedae', *Irish Astronomical Journal* 17 (1985), 98–102

J.A. Bennett, unpublished paper on 'The scientific Community in Nineteenth-Century Ireland', presented at a joint meeting of the Royal Irish Academy and the British Society for the History of Science in 1985 and again to the Cambridge Group for Irish Studies in 1989.

— 'Lord Rosse and the giant reflector', in Nudds et al., *Science in Ireland 1800–1930* (1988), p.105–113

— *Church, State and Astronomy in Ireland: 200 Years of Armagh Observatory* (Armagh/Belfast: Armagh Observatory/Institute of Irish Studies, 1990)

— 'Science and Social Policy in Ireland in the mid-Nineteenth Century', in Bowler and Whyte, *Science and Society in Ireland* (1997), pp. 37–48

H.F. Berry, *A History of the Royal Dublin Society* (London: Longmans, Green & Co., 1915)

J.W. Besant, 'Botanic Gardens: Origin, History and Development', *Journal of the Department of Agriculture* 33 (1935), 173–82

Marie Boas Hall, *All Scientists Now: The Royal Society in the Nineteenth Century* (1984)

Alfred M. Bork, 'The "FitzGerald" Contraction', *Isis* 57 (1966), 199–207

Peter J. Bowler and Nicholas Whyte, eds, *Science and Society in Ireland* (Belfast: Institute of Irish Studies, 1997)

W.H. Brock, N.D. McMillan and R.C. Mollan, eds., *John Tyndall: Essays on a Natural Philosopher* (Dublin: Royal Dublin Society Historical Studies in Irish Science and Technology No. 3, 1981)

Martin Brown, *Hans Sloane* (Blackstaff Primary Science Key Stage 2 textbook; Belfast: Blackstaff Press, 1995)

– *William Traill* (Blackstaff Primary Science Key Stage 2 textbook; Belfast: Blackstaff Press, 1995)

Mary T. Bruck, 'Companions in Astronomy: Margaret Lindsay Huggins and Agnes Mary Clerke', *Irish Astronomical Journal* 20 (1991), 70–7

– 'Agnes Mary Clerke, Chronicler of Astronomy', *Quarterly Journal of the Royal Astronomical Society* 35 (1994), 59–79

Stephen G. Brush, 'Note on the History of the FitzGerald-Lorentz Contraction', *Isis* 58 (1967), 230–32

– 'The Nebular Hypothesis and the Evolutionary Worldview', *History of Science* 25 (1987), 244–69

Patrick Buckland, *Irish Unionism, 1885–1923: A documentary history* (Belfast: HMSO, 1973)

J.E. Burnett and A.D. Morrison-Low, *Vulgar and Mechanick: The Scientific Instrument Trade in Ireland, 1650–1921* (Royal Dublin Society Historical Studies in Irish Science and Technology, no. 8; Edinburgh/Dublin: National Museums of Scotland/RDS, 1989)

C.J. Butler and Ian Elliott, 'Biographical and Historical Notes on the Pioneers of Photometry in Ireland', in *Stellar Photometry: current techniques and future developments (IAU Colloquium 136),* ed. Butler and Elliott (1993)

Anne Buttimer, 'Twilight and Dawn for Geography in Ireland' in Bowler and Whyte, *Science and Society in Ireland* (1997), pp. 135–52

Susan Faye Cannon, *Science in Culture: the Early Victorian Period* (Folkestone: Dawson, 1978)

D.S. Cardwell, *The Organisation of Science in England* (London: Heinemann, 1972)

J.K. Charlesworth, 'Recent progress in Irish geology', *Irish Naturalists' Journal* 6 (1937), 266–74

– 'Recent progress in Irish geology', *Irish Naturalists' Journal* 10 (1950), 61–71

– 'Recent progress in Irish geology', *Irish Naturalists' Journal* 13 (1959), 49–65

– *Historical Geology of Ireland* (Edinburgh and London: Oliver & Boyd, 1963)

– 'Recent Progress in Irish Geology', *Irish Naturalists' Journal* 17 supplement (1972), 1–37

D. Clarke, 'An Outline of the History of Science in Ireland', *Studies* 62 (1973), 287–302

Joe Cleary, 'Irish Culture as Decolonized Nation-State', *Irish Literary Supplement*, Fall 1996, 19–20

M.J. Clune, 'The Work and the Report of the Recess Committee, 1895–1896', *Studies* 71 (1982), 73–84

– 'Horace Plunkett's Resignation from the Irish Department of Agriculture and Technical Instruction, 1906–1907', *Eire-Ireland* 17 (1982), 57–73

Wesley Cocker, 'A History of the University Chemical Laboratory, Trinity College Dublin; 1711–1946', *Hermathena* 124 (1978), 58–76

G.A.J. Cole and T. Hallissy, *Handbook of the Geology of Ireland* (London: Murby & Co, 1924)

R. Collander, *The History of Botany in Finland, 1828–1918,* tr. by David Barrett, (Helsinki: Finnish Academy of Sciences, 1965)

Timothy Collins, *Floreat Hibernia: A Bio-Bibliography of Robert Lloyd Praeger* (Dublin: RDS, 1985)

Tim Pat Coogan, *De Valera: Long Fellow, Long Shadow* (London: Hutchinson, 1993)

Timothy Corcoran SJ, 'The Place of the Sciences in General Education', *Studies* 12 (1923), 406–17

[E.P. Culverwell], *Mr Bryce's Speech on the Proposed Reconstruction of the University of Dublin: Annotated Edition issued by the Dublin University Defence Committee* (Dublin: University Press, 1907)

S.R. Dennison and Oliver MacDonagh, *Guinness 1886–1939: From Incorporation to the Second World War* (Cork: Cork University Press, 1998)

David W. Dewhirst and Michael Hoskin, 'The Rosse Spirals', *Journal for the History of Astronomy* 22 (1991), 257–66

Dictionary of National Biography

Margaret Digby, *Horace Plunkett: An Anglo-American Irishman* (Oxford: Blackwell, 1949)

Henry de Dorlodot, *La Darwinisme au point de vue de l'orthodoxie catholique* (Bruxelles: Vromant, 1921), tr. by Ernest C. Messenger as *Darwinism and Catholic Thought* (London: Burns Oates and Washbourne, 1922)

J.L.E. Dreyer, *A Historical Account of the Armagh Observatory* (Liverpool: H. Greenwood, 1883)

Dublin University Defence Committee, *Trinity College, Dublin, and the Proposed University Legislation for Ireland* (Dublin: University Press, 1907)

Tom Dunne, '*La trahison des clercs:* British intellectuals and the first home-rule crisis', *Irish Historical Studies* 23 (1982), 134–73

G. Elfving, *The History of Mathematics in Finland, 1828–1918* (1981)

Ian Elliott, 'The Monck Plaque', *Irish Astronomical Journal* 18 (1987), 123–4

Havelock Ellis, *A Study of British Genius* (London: Hurst & Blackett, 1904)

T. Enkvist, *The History of Chemistry in Finland, 1828–1918* (Helsinki: Finnish Society of Sciences, 1972)

Samuel Ferguson, *Poems* (Dublin: W. McGee and London: G. Bell, 1880)

David Fitzpatrick, *Politics and Irish Life, 1913–1921: Provincial Experience of War and Revolution* (Dublin: Gill and Macmillan, 1977)

– 'Eamon de Valera at Trinity College', *Hermathena* 133 (1982), 7–14

Georgina Fitzpatrick, *Trinity College and Irish Society 1914–1922* (1992)

John Wilson Foster, 'Natural History, Science and Irish Culture', *The Irish Review* 9 (1990), 61–9

– 'Natural Science and Irish Culture', *Eire-Ireland* 26, no. 2 (1991), 92–103

– 'Natural History in Modern Irish Culture', in Bowler and Whyte, *Science and Society in Ireland* (1997), pp. 119–34

– 'Nature and Nation in the Nineteenth Century', in Foster and Chesney, *Nature in Ireland: A Scientific and Cultural History* (1997), pp. 409–39

– senior editor, and Helena C.G. Chesney, associate editor, *Nature in Ireland: A Scientific and Cultural History* (Dublin: The Lilliput Press, 1997)

R.F. Foster, *Modern Ireland 1600–1972* (London: Penguin, 1988)

Andrew Gailey, 'Unionist Rhetoric and Irish Local Government Reform, 1895–99', *Irish Historical Studies* 24 (1984), 52–68

– *Ireland and the Death of Kindness: The Experience of Constructive Unionism 1890–1905* (Cork: Cork University Press, 1987)

Tom Garvin, 'Priests and Patriots: Irish Separatism and Fear of the Modern, 1890–1914', *Irish Historical Studies* 25 (1986), 67–81

Wilbert Garvin and Des O'Rawe, *Northern Ireland Scientists and Inventors* (Belfast: Blackstaff Press, 1993)

J. Anthony Gaughan, *Alfred O'Rahilly, vol. I: Academic* (Dublin: Kingdom Books, 1986)

H.V. Gill SJ, 'Catholics and Evolution Theories', *Irish Ecclesiastical Record* 19 (1922), 614–24

S.J. Gould, 'Cabinet Museums Revisited', *Natural History* 103 (1994), 12–20

R.P. Graves, *Life of Sir William Rowan Hamilton* (Dublin: Hodges, Figgis and Dublin University Press, 3 vols, 1882, 1885 and 1889)

T.L. Hankins, *Sir William Rowan Hamilton* (Baltimore and London: Johns Hopkins Press, 1980)

Michael Hechter, *Internal Colonialism: The Celtic Fringe in British National Development, 1536–1966* (1975)

Gordon L. Herries Davies, 'The Earth Sciences in Irish Serial Publications 1787–1977', *Journal of Earth Sciences, Royal Dublin Society* 1 (1978), 1–23

– *Sheets of Many Colours: The Mapping of Ireland's Rocks, 1750–1890* (Royal Dublin Society Historical Studies in Irish Science and Technology, no. 4; Dublin: RDS, 1983)

- 'Astronomy, Geology, Meteorology', in T. O Raifeartaigh, ed., *The Royal Irish Academy: A Bicentennial History, 1785–1985* (1985)
- 'The History of Irish Science: A Select Bibliography, second edition' (duplicated typescript; Dublin: Royal Irish Academy, 1985)
- *North from the Hook: 150 Years of the Geological Survey of Ireland* ([Dublin:] Geological Survey of Ireland, 1995)
- 'Irish Thought in Science', in *The Irish Mind,* ed. Richard Kearney (Dublin: Wolfhound, 1985)

D. Hoctor, *The Department's Story – A History of the Department of Agriculture* (Dublin: Institute of Public Administration, 1971)

Michael Hoskin, 'The First Drawing of a Spiral Nebula', *Journal for the History of Astronomy* 13 (1982), 97–101
- 'Archives of Dunsink and Markree Observatories', *Journal for the History of Astronomy* 13 (1982), 149–52
- 'Rosse, Robinson and the Resolution of the Nebulae', *Journal for the History of Astronomy* 21 (1990), 331–44

E. Hull, *Reminiscences of a Strenuous Life* (London: H. Rees, 1910)

Bruce J. Hunt, 'The Origins of the FitzGerald Contraction', *British Journal of the History of Science* 21 (1988), 67–76

J. de C. Ireland, *Ireland's Sea Fisheries: A History* (Dun Laoghaire: Glendale, 1981)

A. Jackson, 'Irish Unionism and the Russellite Threat, 1894–1906', *Irish Historical Studies* 25 (1987), 376–404
- 'The Failure of Unionism in Dublin, 1900', *Irish Historical Studies* 26 (1989), 377–95

Richard Jarrell, 'The Department of Science and Art and Control of Irish Science, 1853–1905', *Irish Historical Studies* 23 (1983), 330–47
- 'Differential National Development and Science in the Nineteenth Century: The Problems of Quebec and Ireland', in *Scientific Colonialism,* ed. Reingold and Rothenberg (1987)

Roy Johnston, 'Science and Technology in Irish National Culture', *Crane Bag* 7 (1983), 58–63
- 'Science in a Post-Colonial Culture', *Irish Review* 8 (1990), 70–6
- 'Godless Colleges and Non-Persons', *Causeway* 1, no. 1 (Autumn 1993), 36–8

[John Joly], 'In Trinity College During the Sinn Féin Rebellion, by One of the Garrison', *Blackwood's Magazine* 200 (July 1916), 101–25.

Greta Jones, 'Eugenics in Ireland – the Belfast Eugenics Society, 1911–1915', *Irish Historical Studies* 28 (1992), 81–95
- 'Science, Catholicism and Nationalism', *Irish Review* 20 (1997), 47–61
- and Elizabeth Malcolm, eds, *Medicine, Disease and the State in Ireland, 1650–1940* (Cork: Cork University Press, 1998)

Jonas Jørstad, 'Nations once again', in *Revolution? Ireland 1917–1923*, ed. David Fitzpatrick (Dublin: Trinity History Workshop, 1990)

Eino Jutikkala and Kauko Pirinen, *A History of Finland* (London: Heinemann, 1979)

B.B. Kelham, 'The Royal College of Science for Ireland (1867–1926)', *Studies* 56 (1967), 297–309

Liam Kennedy, *Colonialism, Religion and Nationalism in Ireland* (Belfast: Institute of Irish Studies, 1996)

– 'Modern Ireland: Post-Colonial Society or Post-Colonial Pretensions?' *The Irish Review* 13 (Winter 1992/93), 107–21

Peggy Aldrich Kidwell, 'Women Astronomers in Britain, 1780–1930', *Isis* 75 (1984), 534–46

E.T. Kinzer and J. Fukai, 'Weber's Force and Maxwell's Equations', *Foundations of Physics Letters* 9 (1996), 457–61

H. Kragh, *An Introduction to the Historiography of Science* (Cambridge: CUP, 1987)

V.V. Krishna, 'The Colonial "Model" and the Emergence of National Science in India: 1876–1920', in P. Petitjean et al., eds, *Science and Empires* (1992)

H. Kucklick, *The Savage Within: The Social History of British Anthropology, 1885–1945* (Cambridge: CUP, 1991)

John Lankford, 'Amateurs and Astrophysics: A Neglected Aspect in the Development of a Scientific Specialty', *Social Studies of Science* 11 (1981), 275–303

Bruno Latour, *The Pasteurization of France* (Cambridge & London: Harvard University Press, 1988)

J.J. Lee, *Ireland 1912–1985: Politics and Society* (Cambridge: CUP, 1989)

David N. Livingstone, 'Darwin in Belfast: The Evolution Debate', in Foster and Chesney, *Nature in Ireland* (1998), pp. 387–408

Lord Longford, *Peace by Ordeal* (London: Sidgwick and Jackson, 1972)

J. Loughlin, *Gladstone, Home Rule and the Ulster Question, 1882–1893* (Dublin: Gill & Macmillan 1986)

– 'T.W. Russell, the Tenant-Farmer Interest, and Progressive Unionism in Ulster, 1886–1900', *Eire-Ireland* 25 (1990), 44–63

A.M. Lucas, P.J. Lucas, T.A. Darragh & S. Maroske, 'Colonial Pride and Metropolitan Expectations: The British Museum and Melbourne's Meteorites', *British Journal for the History of Science* 27 (1994), 65–87

A.A. Lynch, *Ireland: Vital Hour* (London: S. Paul, 1915)

– *Principles of Psychology; the Foundation Work of the Alethian System of Philosophy* (London: G. Bell, 1923)

– *The Case against Einstein* (London: Allan, 1932)

Sean Lysaght, 'Heaney vs. Praeger: Contrasting Natures', *Irish Review* 7 (1989), 68–74

- 'Robert Lloyd Praeger and the Culture of Science in Ireland: 1865-1953' (Ph.D. thesis, NUI 1994)
- 'Themes in the Irish History of Science', *Irish Review* 19 (1996), 87-97
- 'Science and the Cultural Revival, 1863-1916', in Bowler and Whyte, *Science and Society in Ireland* (1997), pp. 153-66
- 'Contrasting Natures: The Issue of Names', in Foster and Chesney, *Nature in Ireland* (1997), pp. 440-60

Lawrence W. MacBride, *The Greening of Dublin Castle: The Transformation of Bureaucratic and Judicial Personnel in Ireland, 1892-1922* (Washington: Catholic University of America Press, 1991),

Michael McCarthy, *Five Years in Ireland: 1895-1900* (London: Simpkin, Marshall & Co., and Dublin: Hodges, Figgis, 1901)

Donal McCartney, *The National University of Ireland and Eamon de Valera* (Dublin: University Press of Ireland, 1983)

John McColgan, *British Policy and the Irish Administration, 1920-22* (London: Allen & Unwin, 1983)

W. McDonald, *On Motion* (Dublin: Browne and Nolan, 1898)
- *Reminiscences of a Maynooth Professor* (London: Jonathan Cape, 1920; second edition, Cork: Mercier Press, 1967)

R.B. McDowell, *The Irish Convention 1917-1918* (London: Routledge & Kegan Paul, 1970)
- 'The Main Narrative', in T. O Raifeartaigh, ed., *The Royal Irish Academy: A Bicentennial History, 1785-1985* (Dublin: RIA, 1985)
- and D.A. Webb, *Trinity College Dublin, 1592-1952: An Academic History* (Cambridge: CUP, 1982)

Fergal McGrath, *Newman's University: Idea and Reality* (London, etc.: Longmans, 1951)
- 'The University Question', third section of *A History of Irish Catholicism v. 5/6* ([Dublin:] Gill & Macmillan, [1971])

Noeleen McHenry and Maurice McHenry, *John Clarke* (Blackstaff Primary Science Key Stage 2 textbook; Belfast: Blackstaff Press, 1995)

Susan M.P. McKenna, 'Astronomy in Ireland from 1780', *Vistas in Astronomy* 9 (1968), 283-96

Susan McKenna-Lawlor, 'Astronomy in Ireland from the Late Eighteenth to the End of the Nineteenth Century', in Nudds et al., *Science in Ireland 1800-1930* (1988), pp. 85-96.
- *Whatever Shines Should be Observed [quicquid nited notandum]* (Samton Historical Studies No. 3; Blackrock, Co. Dublin: Samton Ltd, 1998)
- and Michael Hoskin, 'Correspondence of Markree Observatory', *Journal for the History of Astronomy* 15 (1984)

P.A. McKeown, 'T.W. Russell: Temperance Agitator, Militant Unionist Missionary, Radical Reformer and Political Pragmatist' (Ph. D. thesis, QUB, 1991)

P.J. McLaughlin, 'A Century of Science in the *Irish Ecclesiastical Record*: Monsignor Molloy and Father Gill', *Irish Ecclesiastical Record*, 5th ser., 101 (Jan. 1964), 34–40

Roy MacLeod, 'Reflections on the Architecture of Imperial Science', in *Scientific Colonialism*, ed. Reingold and Rothenberg (1987), pp. 217–49

– 'On Science and Colonialism', in Bowler and Whyte, *Science and Society in Ireland* (1997), pp. 1–18

E. MacLysaght, *Sir Horace Plunkett and his Place in the Irish Nation* (London & Dublin: Maunsel, 1916)

N.D. MacMillan, 'Organisation and Achievements of Irish Astronomy in the Nineteenth Century – Evidence for a "Network"', *Irish Astronomical Journal* 19 (1990), 101–18

Derek McNally and Michael Hoskin, 'William E. Wilson's Observatory at Daramona House', *Journal for the History of Astronomy* 19 (1988), 146–53

Louis McRedmond, *Thrown Among Strangers: John Henry Newman in Ireland* (Dublin: Veritas, 1990)

J. Mauchline, D.J. Ellett, J.D. Gage, J.D.M. Gordon and E.J.W. Jones, 'A Bibliography of the Rockall Trough', *Proceedings of the Royal Society of Edinburgh, Section B – Biological Sciences* 88 (1986), 319–54.

James Meenan and Desmond Clarke, eds, *RDS: The Royal Dublin Society, 1731–1981* (Dublin: Gill & Macmillan, 1981)

Robert Merton, 'Science, Technology and Society in Seventeenth Century England', *Osiris* 4 (1938), 360–632

Arthur Mitchell, *Revolutionary Government in Ireland: Dáil Éireann, 1919–22* (Dublin: Gill & Macmillan, 1995)

Charles Mollan, Eileen Byrne, Judy Bright, Nuala O'Duffy and Brid Desmond, eds, *Nostri Plena Laboris: An Author Index to the RDS Scientific Journals, 1800–1965*

Charles Mollan, William Davis and Brendan Finucane, eds, *Some People and Places in Irish Science and Technology* (Dublin: RIA, 1985)

Charles Mollan, William Davis and Brendan Finucane, eds, *More People and Places in Irish Science and Technology* (Dublin: RIA, 1990)

Gerald Molloy, *Geology and Revelation: or the Ancient History of the Earth Considered in the Light of Geological Facts and Revealed Religion* (London: Longmans, Green, Reader, and Dyer, 1870)

Michael Francis Moloney, *Irish Ethno-Botany and the Evolution of Medicine in Ireland* (Dublin: M.H. Gill, 1919)

N.T. Monaghan, 'Geology in the National Museum of Ireland', *Geological Curator* 5 (1992 for 1989), 275–82

Sally Montgomery, *Jocelyn Bell Burnell* (Blackstaff Primary Science Key Stage 2 textbook; Belfast: Blackstaff Press, 1995)

– *Robert Lloyd Praeger* (Blackstaff Primary Science Key Stage 2 textbook; Belfast: Blackstaff Press, 1995)

Christopher Moriarty, 'Fish and Fisheries', in Foster and Chesney, *Nature in Ireland* (1997), pp. 283–98

T.J. Morrissey, *Towards a National University: William Delany S.J. (1835–1924) – An Era of Initiative in Irish Education* (Dublin: Wolfhound, 1983)

T.W. Moody, 'The Irish University Question of the Nineteenth Century', *History* 43 (1958), 90–109

George Moore, *Hail and Farewell: Ave, Salve, Vale,* edited with notes by Richard Cave (Toronto: Macmillan, 1976)

James R. Moore, *The Post-Darwinian Controversies: A Study of the Protestant Struggle to Come to Terms with Darwin in Great Britain and America, 1870–1900* (Cambridge: CUP, 1979).

Patrick Moore, *Armagh Observatory: A History, 1790–1967* (Armagh: Armagh Observatory, 1967)

– *The Astronomy of Birr Castle* (London: Mitchell Beazley, 1971)

D.P. Moran, *The Philosophy of Irish Ireland* (Dublin: Duffy, 1905)

Jack Morrell and Arnold Thackray, *Gentlemen of Science: Early Years of the British Association for the Advancement of Science* (London: Offices of the Royal Historical Society, University College London, 1981)

Anne-Marie Moulin, 'Patriarchal Science: The Network of the Overseas Pasteur Institutes', in *Science and Empires*, ed. Patrick Petitjean et al. (1992)

J.A. Murphy, *The College: A History of Queen's/University College Cork, 1845–1995* (Cork: Cork University Press, 1995)

Robert H. Murray, *Archbishop Bernard: Professor, Prelate and Provost* (London: SPCK, 1931)

E. Charles Nelson, *The Brightest Jewel: A History of the National Botanic Gardens, Glasnevin, Dublin* (Kilkenny: Boethius, 1987)

T.E. Nevin, 'Experimental Physics', in T. O Raifeartaigh, ed., *The Royal Irish Academy: A Bicentennial History, 1785–1985* (1985)

John Henry Newman, *The Idea of a University, Defined and Illustrated* (London: Pickering, 1873)

J.R. Nudds, N.D. McMillan, D.L. Weaire and S.M.P. McKenna-Lawlor, eds, *Science in Ireland, 1800–1930, Tradition and Reform, Proceedings of an International Symposium Held at Trinity College, Dublin, March, 1988* (Dublin: TCD, 1988)

J.R. Nudds, 'John Joly: Brilliant Polymath', in Nudds et al., *Science in Ireland 1800–1930* (1988), pp. 163–78

C. Cruise O'Brien, *Parnell and his Party, 1880–90* (Oxford: Clarendon, 1957)

James M. O'Connor, 'Insects and Entomology', in Foster, *Nature in Ireland* (1998), pp. 219–40

S. O'Donnell, *William Rowan Hamilton – Portrait of a Prodigy* (Dublin: Boole Press, 1983)

James L. O'Donovan, 'Experimental Research in University College, Dublin', *Studies* 10 (1921), 109–22

Eunan O'Halpin, *The Decline of the Union: British Government in Ireland, 1892–1920* (Dublin: Gill & Macmillan, 1987)

C. O hEocha, 'Biology', in T. O Raifeartaigh, ed., *The Royal Irish Academy: A Bicentennial History, 1785–1985* (1985)

Robert Olby, 'Social Imperialism and State Support for Agricultural Research in Edwardian Britain', *Annals of Science* 48 (1991), 509–26

Ollsgoil na h-Eireann: The National University Handbook, 1908–1932 (Dublin: NUI, 1932)

Alfred O'Rahilly, *Electromagnetic Theory: A Critical Examination of Fundamentals* (reprinted New York: Dover, 1965)

Ronan O'Rahilly, *The Cork Medical School, 1849–1949* (Cork: Cork University Press, 1949)

C.E. O'Riordan, *The Natural History Museum, Dublin* (Dublin: Stationery Office, c.1983)

– *A Catalogue of the Collection of Irish Marine Crustacea in the National Museum of Ireland* (Dublin: Stationery Office, c.1969)

T. O Raifeartaigh, ed., *The Royal Irish Academy: A Bicentennial History, 1785–1985* (Dublin: RIA, 1985)

Katherine O'Shea (Mrs Charles Stewart Parnell), *Charles Stewart Parnell: His Love Story and Political Life* (London: Cassell, 1914), 2 vols

D. O'Sullivan, *The Irish Free State and its Senate: A Study in Contemporary Politics* (London: Faber & Faber, 1940)

Dorinda Outram, 'Negating the Natural: Or Why Historians Deny Irish Science', *The Irish Review* 1 (1986), 45–9

'Student', a Statistical Biography of William Sealy Gosset, based on writings by E.S. Pearson, edited and augmented by R.L. Plackett with the assistance of G.A. Barnard (Oxford: Clarendon, 1990)

Patrick Petitjean, Catherine Jami and Anne-Marie Moulin, eds, *Science and Empires: Historical Studies about Scientific Development and European Expansion* (Boston Studies in the Philosophy of Science, no. 136; Dordrecht: Kluwer Academic Publishers, 1992)

E.M. Philbin, 'Chemistry', in T. O Raifeartaigh, ed., *The Royal Irish Academy: A Bicentennial History, 1785–1985* (Dublin: RIA, 1985)

Sheila Pim, *The Wood and the Trees* (2nd edn; Kilkenny: Boethius, 1984)

Sir Horace Plunkett, *Ireland in the New Century* (3rd edition; New York: Dutton, 1908)

Robert Lloyd Praeger, *The Way that I Went: An Irishman in Ireland* (Dublin: Hodges Figgis, 1937)

– *A Populous Solitude* (Dublin: Hodges Figgis, 1941)

– *Some Irish Naturalists* (Dundalk: W. Tempest, 1949)

H. Price, *Fifty Years of Psychical Research: A Critical Survey* (London: Longmans, 1939)

D.J. [de Solla] Price, 'The exponential curve of science', *Discovery* 17 (1956), 240–3

A. Prochaska, *Irish History from 1700: A Guide to Sources in the Public Record Office* ([London]: British Records Association, 1986)

Pierce F. Purcell, *The Peat Resources of Ireland: A Lecture Given before the Royal Dublin Society on 5th March 1919* (London: DSIR, Fuel Research Board Special Report No. 2, 1920)

A. Hingston Quiggin, *Haddon the Head-Hunter* (Cambridge: CUP, 1942)

Recess Committee, *Report by the Recess Committee on the Establishment of a Department of Agriculture and Industries for Ireland* (Dublin: Browne and Nolan, 1896, 2nd edn 1906)

Nathan Reingold and Marc Rothenberg, eds, *Scientific Colonialism: A Cross-Cultural Comparison (Papers from a Conference at Melbourne, Australia, 25–30 May 1981)* (Washington: Smithsonian Institution Press, 1987)

J.D. Root, 'The Final Apostasy of St George Jackson Mivart', *Catholic Historical Review* 71 (1985), 1

[3rd] Earl of Rosse, *Letters on the State of Ireland* (London: Hatchard, 1847)

Royal Commission on the University of Dublin, *Report*, 1920 (Cmd. 1078)

Royal Dublin Society, *Bicentenary Souvenir* (Dublin: Browne & Nolan, 1931)

T.W. Russell, *Ireland and the Empire* (London: G. Richards, 1901)

Simon Schaffer, 'The Nebular Hypothesis and the Science of Progress', in *History, Humanity and Evolution: Essays for John C. Greene*, ed. James R. Moore (Cambridge: CUP, 1989), pp. 131–64

R.F. Scharff, *European Animals: Their Geological History and Geographical Distribution* (London: A. Constable, 1907)

S. Shapin and A. Thackray, 'Prosopography as a Research Tool in History of Science: The British Scientific Community 1700–1900', *History of Science* 12 (1974), 1–28

Susan Sheets-Pyenson, *Cathedrals of Science: The Development of Colonial Natural History Museums during the Late Nineteenth Century* (Kingston/Montreal: McGill/Queen's University Press, 1988)

Fathers of the Society of Jesus, *A Page of Irish History: Story of University College, Dublin, 1883–1909* (Dublin: Talbot, 1930)

H. Soodak and M.S. Tierstein, 'Dynamic Interpretation of Maxwell's Equations', *American Journal of Physics* 64 (1994), 907–13

D. Taylor and M. McGuckian, 'The Three-Foot Telescope at Birr and Lunar Heat', in Nudds et al., *Science in Ireland 1800–1930* (1988), pp. 115–22

M. Tierney, ed., *Struggle with Fortune: A Miscellany for the Centenary of the Catholic University of Ireland, 1854–1954* (Dublin: Browne & Nolan, [1954])

E. Brian Titley, *Church, State and the Control of Schooling in Ireland, 1900–1914* (Dublin: Gill & Macmillan, 1983)

John Tyndall, *Address Delivered before the British Association Assembled at Belfast, with Additions* (London: Longmans, 1874)

— *Fragments of Science, vol. II* (6th edn; London: Longmans, 1879)

T.D. Spearman, 'Mathematics and Theoretical Physics', in T. O Raifeartaigh, ed., *The Royal Irish Academy: A Bicentennial History, 1785–1985*

G. Johnstone Stoney, 'On the Cause of Double Lines and of Equidistant Satellites in the Spectra of Gases', *Scientific Transactions of the Royal Dublin Society* 4 (1891), 563–608

William Irwin Thompson, *The Imagination of an Insurrection: Dublin, Easter, 1916* (Oxford: Oxford University Press, 1967)

Maryann Giulanella Valiulis, *Portrait of a Revolutionary: General Richard Mulcahy and the Founding of the Irish Free State* (Blackrock, Co. Dublin: Irish Academic Press, 1992)

K[laas] van Berkel, *In het Voetspoor van Stevin: Geschiedenis van de Natuurwetenschap in Nederland 1580–1940* (Meppel: Boom, 1985)

Michael Viney, 'Woodcock for a Farthing: The Irish Experience of Nature', *Irish Review* 1 (1986), 58–64

Brian M. Walker, *Parliamentary Election Results in Ireland, 1918–92* (Dublin/Belfast: Royal Irish Academy/Institute for Irish Studies, 1992)

P.A. Wayman, *Dunsink Observatory, 1785–1985: A Bicentennial History* (Royal Dublin Society Historical Studies in Irish Science and Technology, no. 7; Dublin: DIAS & RDS, 1987)

— 'The Grubb Astrographic Telescopes, 1887–1896', in *Mapping the Sky: Past Heritage and Future Directions (IAU Symposium 133)*, ed. S. Debarbat et al. (1988), pp. 139–42

Trevor West, *Horace Plunkett, Co-Operation and Politics: An Irish Biography* (Gerrards Cross: Colin Smythe, 1986)

Terence de Vere White, *Kevin O'Higgins* (London: Methuen, 1948)

— *The Story of the Royal Dublin Society* (Tralee: The Kerryman, 1955)

J.H. Whyte, *Interpreting Northern Ireland* (Oxford: University Press, 1990)

Nicholas Whyte, '"Lords of Ether and of Light": The Irish Astronomical Tradition of the Nineteenth Century', *The Irish Review* 17/18 (1995), 127–41

— 'Digging up the Past: A Visit to Markree Castle', *Stardust: The Journal of the Irish Astronomical Association* 25/4 (1992), 27–30

— 'Science and Nationality in Edwardian Ireland', in Bowler and Whyte, *Science and Society in Ireland* (1997)

B.A.C. Windle, *The Church and Science* (London: Catholic Truth Society, 1918)

W. Fitzgerald Woodworth and M.J. Gorman, *The College of Science for Ireland: Its Origin and Development* (Dublin: University Press, 1923)

Patrick Wyse-Jackson, 'Fluctuations in Fortune: Three Hundred Years of Irish Geology', in Foster, *Nature in Ireland* (1998), pp. 91–114

Steve Yearley, 'Colonial Science and Dependent Development: The Case of the Irish Experience', *Sociological Review* 37 (1989)

B: Periodicals

Hansard, *Parliamentary Debates*
Dáil Debates
Senate Debates
Irish Times
Freeman's Journal
Belfast Newsletter
The Torch (in National Library of Ireland)
Annual Reports, Department of Agriculture and Technical Instruction
Annual Reports, Department of Scientific and Industrial Research
Who's Who

C: Archival sources

British Geological Survey, Keyworth, Nottinghamshire: Geological Survey archives

British Museum (Natural History), Kensington: Crustacea and Keeper's correspondence

Cambridge University Library: Kelvin, Stokes, Larmor, and Bernal papers

Geological Survey of Ireland, Dublin

National Archives, Dublin: DATI archives

National Library of Ireland: T.P. Gill papers

National Museum, Dublin: Director's correspondence, Natural History correspondence

Oxford University, Bodleian Library: British Association for the Advancement of Science

Plunkett Foundation, Oxford: Horace Plunkett's correspondence and diaries

Public Records Office, Kew

Royal Dublin Society: Minute Books, Committee for Science and its Industrial Applications

Royal Irish Academy: Minutes of the Council of the Academy between 1919 and 1922

Trinity College, Dublin: Bernard papers, Joly papers

University College, Dublin: Corcoran papers, Royal College of Science archives

University College, London: Oliver Lodge correspondence

Index